SpringerBriefs in Philosophy

For further volumes:
http://www.springer.com/series/10082

Peter Truran

Practical Applications of the Philosophy of Science

Thinking About Research

 Springer

Peter Truran
Center for the Philosophy of Science
University of Minnesota
Minneapolis, MN
USA

ISSN 2211-4548   ISSN 2211-4556   (electronic)
ISBN 978-3-319-00451-8   ISBN 978-3-319-00452-5   (eBook)
DOI 10.1007/978-3-319-00452-5
Springer Cham Heidelberg New York Dordrecht London

Library of Congress Control Number: 2013938749

© The Author(s) 2013
This work is subject to copyright. All rights are reserved by the Publisher, whether the whole or part of the material is concerned, specifically the rights of translation, reprinting, reuse of illustrations, recitation, broadcasting, reproduction on microfilms or in any other physical way, and transmission or information storage and retrieval, electronic adaptation, computer software, or by similar or dissimilar methodology now known or hereafter developed. Exempted from this legal reservation are brief excerpts in connection with reviews or scholarly analysis or material supplied specifically for the purpose of being entered and executed on a computer system, for exclusive use by the purchaser of the work. Duplication of this publication or parts thereof is permitted only under the provisions of the Copyright Law of the Publisher's location, in its current version, and permission for use must always be obtained from Springer. Permissions for use may be obtained through RightsLink at the Copyright Clearance Center. Violations are liable to prosecution under the respective Copyright Law.
The use of general descriptive names, registered names, trademarks, service marks, etc. in this publication does not imply, even in the absence of a specific statement, that such names are exempt from the relevant protective laws and regulations and therefore free for general use.
While the advice and information in this book are believed to be true and accurate at the date of publication, neither the authors nor the editors nor the publisher can accept any legal responsibility for any errors or omissions that may be made. The publisher makes no warranty, express or implied, with respect to the material contained herein.

Printed on acid-free paper

Springer is part of Springer Science+Business Media (www.springer.com)

Preface

I developed an interest in the philosophy of science some years after the completion of my undergraduate studies in biochemistry. The origins of this book lie in my recognition that some understanding of the philosophy of science would be helpful with teaching high school biology students how to carry out laboratory and field studies in ways which truly reflected the research process. Subsequently, this newly acquired knowledge helped propel me back into the world of research, a doctoral degree, and ultimately into a career as a development scientist in the public sector.

During the course of my professional life I tried to convey these ideas to others, and referred them to what I considered to be useful introductions to the philosophy of science. While there were plenty of manuals describing how to carry out a research project, there seemed to be no simple, accessible guide showing how the philosophy of science can inform research practice.

This book is intended to fill this gap, by encouraging undergraduate and graduate students to achieve an understanding of the philosophy of science that provides practical insights into the attitudes and methods necessary for a successful approach to research. I hope it will also be useful to their professors and advisors, and that it will engender an enduring interest in the philosophy of science.

Acknowledgments

I am indebted to Prof. C. Kenneth Waters, Director of the Minnesota Center for the Philosophy of Science, University of Minnesota, for inviting me to be a visiting fellow at the Center for an extended period. Prof. Waters offered generous support with resources and encouragement, and the opportunity to be included in the intellectual life of the Center.

I gratefully acknowledge the support of Geoffrey Gorham, Macalester College, St Paul and Alan Love, University of Minnesota, Minneapolis, who provided invaluable critical review. I also wish to thank the members of the Biological Interest Group who, week by week, contribute to an ongoing discussion of the relationship between the philosophy and history of science, and research in biology. This inclusive and pluralistic forum has been instrumental in my philosophical education.

Stephen Peters provided much needed critique and editorial advice from the beginning. Finally, I wish to express my deep appreciation to Sylvia Horwitz for providing inspiration and patient support throughout the writing of this book.

Minneapolis, MN   Peter Truran

Contents

Chapter 1
Introduction: A Personal Perspective

Abstract The philosophy of science, when it speaks directly to scientists, provides inspiration and guidance as to the best state of mind required to approach the process of scientific inquiry. It can provide guidance to those embarking on a research career and those entrusted with the education of young scientists.

Science is a messy business. When historians and philosophers of science look at the ways in which scientists do research, they find that science progresses by strokes of genius and by dogged application, with twists and turns and dead ends. Discoveries may involve serendipity or even dumb luck. The vast community of scientists includes those doing groundbreaking science and those doing mundane tasks; it includes some bad scientists and even a few crooks. Those who have the talent and good fortune to make a major discovery oblige the rest of us to see the world in a new way, to bring about what Thomas Kuhn would call a "paradigm shift" in our thinking. Others do the solid "normal science" that is the lot of most researchers—those who may not achieve special public recognition but who make essential contributions to the body of scientific knowledge.

One of the tasks of the philosophy of science is to tease out and describe the ways in which good science is done, and to distinguish it from non-science (metaphysics), from pseudoscience (non-science posing as science), and from bad science. It attempts to provide an understanding of scientific research that is based upon studies of historical and contemporary practice, and is shaped by a continuing dialogue with working scientists. However, in addition to offering a descriptive account of science, it may also constitute a view of how science *ought* to be practiced—a normative account. As such, it can provide guidance to those embarking on a research career and those entrusted with the education of young scientists. The philosophy of science, when it speaks directly to scientists, provides inspiration and guidance as to the best state of mind required to approach scientific inquiry. It allows the deployment of logical frameworks that inform research practice, and it gives insight into the processes of problem solving. When it speaks to

P. Truran, *Practical Applications of the Philosophy of Science*,
SpringerBriefs in Philosophy, DOI: 10.1007/978-3-319-00452-5_1,
© The Author(s) 2013

non-scientists, it helps to break down the walls that have been erected between the cultures of the arts and the sciences, showing that the ways of thinking used in mainstream science can be useful in tackling problems in areas where the application of scientific method has traditionally not been considered appropriate or possible. More recent developments in the philosophy of science reflect on the ways in which scientists behave as a community of researchers and show how scientists engage with society as a whole.

At the core is the need to understand clearly that science is not just a body of knowledge, but is a process of inquiry. Philosophers of science attempt to articulate the ways in which this process is conducted and, in particular, the relationship between observations and theory, the theory-laden nature of observations, and the role of experiment. They investigate the nature of scientific knowledge, its relationship to the world beyond appearances, and its status with respect to other forms of knowledge. They help scientists to recognize and defend science against the claims of pseudoscience.

There is no single way of doing science. Within the philosophy of science it is recognized that there are different problems and methodologies associated with the practice of different scientific disciplines. The philosophy of biology, for example, has its own range of concerns, the debates around evolution and creationism, or behavioral genetics, which have no counterpart in physics. The leading edges of the philosophy of science exhibit a high degree of pluralism, with lively debates about the best means for conceiving issues and addressing problems. However, some these may be highly technical or abstruse, and many of the specialist concerns of professional philosophers may seem to have little relevance for researchers. Nevertheless, there are points where the day-to-day activities of a working scientist may be informed by the philosophy of science so as to confer tangible benefits upon his research practice. Looking back over a career spanning high school teaching, research, and working in the public sector, I can see how my encounter with the philosophy of science shaped my views of how science should be taught and practiced.

As a student, my understanding of science reflected the didactic nature of the sources of my knowledge. I developed some knowledge of the content of science, but I had little awareness of the processes by which the discoveries I read about had been made. My experience of scientific method in the high-school classroom environment was similar to that of most school students of my generation—experiments that were really demonstrations, or investigations that were really laboratory exercises designed to get "the right result." This emphasis on content rather than method continued throughout my undergraduate studies in biology and biochemistry. In the laboratory we cut up and made detailed drawings of plants and animals, learned to use a microscope, and followed recipes for the extraction and characterization of enzymes. There was little consideration of the role of experimental design and statistics within the biological sciences. Eventually, in my final year, I completed a small research project that, in retrospect, reflected my complete ignorance of scientific method. I produced a set of uniformly negative results with no understanding of whether they were a true reflection of the phenomenon under study, or due to the inadequacy of the measurement system being used. I ended my university education with a feeling of disenchantment with science.

I first developed an interest in the philosophy of science when working as a high school biology teacher several years after completing my undergraduate studies. I began to understand, for the first time, the ways in which science makes progress. My understanding informed the ways in which I approached my teaching, especially in the area of laboratory-based exercises and investigations. I saw that the students' understanding of the principles of scientific methodology provided them with a powerful motivation for carrying out experimental work in a careful and purposeful manner. It gave them a sense of what it is to work as a scientist, developing ideas and testing them rigorously. This was a far cry from the experience of following a teacher's instructions by rote in order to achieve a result that was already known by the teacher. This approach enabled me to become a more effective science teacher and it had the side effect of re-kindling my interest in pursuing a career in scientific research. Subsequently, when I resumed my scientific career as a researcher in endocrinology at the University of Wales, I found that my understanding of the philosophy of science provided valuable insights into my doctoral work. It helped me to understand the importance of clearly framed hypotheses, and the development of experimental designs and statistical analyses that took account of predicted results.

Looking back at this experience of learning how to do research in a busy and productive scientific institution, it is clear that what I was doing could be characterized as a process of apprenticeship. More than anything else, this process was shaped by the relationship with my advisor, the "master scientist" who guided, encouraged, and directed me. Much of what I learned was practical. I watched and listened, and received instruction from experienced technicians. They initiated me into the disciplines of good laboratory practice and the skills required for the manipulation of small volumes of patient samples, reagents, and radioactive materials. Attending scientific meetings helped me to develop an awareness of the unwritten rules of etiquette governing the proceedings. I learned how to write and present research papers, how to submit them to journals for publication, and how to respond to the criticism of editors and referees. I engaged in dialogue with other scientists and experienced the process of generating of ideas and hypotheses in tearooms, in meetings, and at the bench. Most important of all was learning about the standards by which the quality and integrity of scientific work were judged.

However, there was very little discussion of what we were doing in philosophical terms. Like me, many of my fellow researchers had experienced very little exposure to the philosophy of science during their education. Those who had were wary of its relevance to the practicalities of research, and I soon learned not to introduce philosophical concepts into my conversations with them. I still continued to read about the philosophy of science, and to reflect on its application to my work as a researcher. As I became more confident about my own understanding of what it meant to be a practicing scientist, I realized that the ways that my fellow apprentice scientists learned their trade would be improved by their having an understanding of the philosophy of science. Indeed, it was my impression that the work of many of the experienced scientists I encountered would also benefit from such an understanding, particularly when it came to the process of developing in their "apprentices" the critical attitudes and understandings required to become effective scientists. It was at that point that the idea of writing an accessible introduction to the philosophy of science first occurred to me.

After a number of years spent honing my skills in post-doctoral research, I moved on to a career in the private sector and spent this greater part of my working life supporting the development and maintenance of new immunodiagnostic products—automated systems for the rapid measurement of concentrations of hormones, disease antibodies and antigens, and other clinically important biological markers circulating in the bloodstream. This was a highly pressurized world with no place for sentiment or idle reflection.

Nevertheless, I found that my knowledge of the philosophy of science gave me an edge in terms of my effectiveness as a working scientist and manager. The research and development of new products was done under the pressure of aggressive schedules, and problems with launched products had to be resolved while maintaining the continuity of supply to the customer. Under these circumstances, it was tempting to start gathering easily available data to support the confirmation of ideas that had arisen from hunches or from "brain-storming" sessions. Ultimately it was much more effective to use statistically based experimental designs with testable hypotheses. Well thought-out designs were more likely to yield unambiguous results that, even if negative, allowed clear decisions to be made.

Much of my time as a scientist/manager was spent protecting young professional scientists from these pressures while initiating them into the disciplines of scientific problem solving. Many of them lacked the conceptual framework required for a meaningful discussion of scientific method. I was struck by the impact made when introducing them to philosophical ideas of rationality, the need for an open but critical mind, the theory-laden nature of observations, and the need for the clear formulation of testable hypotheses. By the time I came to the end of my career in the public sector, my ideas were beginning to crystallize around writing a primer on the philosophy of science that would make clear its practical benefits for research. It would provide guidance to those immersed in the process of scientific apprenticeship and to those who were supervising them.

It is not easy to convince scientists of the relevance of the philosophy of science to research practice. Much of this skepticism may arise from the ways in which the philosophy of science is perceived to employ arcane concepts and terminology, and to bear only a peripheral relationship to the real world of research. Working scientists tend to be intensely pragmatic, problem oriented and wary of obscurantism. Accordingly, this account of the philosophy of science is structured around the problems likely to be encountered in research: the choice of a feasible problem, the relationships between theory and observation, experimental design and statistics, and the cultivation of a scientific attitude. It aims to provide conceptual tools to facilitate the dialogue between philosophers and scientists, to allow the scientist to conduct a defense against the claims of pseudoscience and anti-scientific skepticism, and to engage with those who, while not scientists themselves, may have responsibilities for policy and funding decisions which may affect research programs. It adopts the position of current philosophy of science that offers no single definitive view of the way in which science makes progress. Rather it attempts to provide a non-dogmatic account that remains accessible without compromising rigor.

An understanding of the philosophy of science can make a real difference to the ways we approach the problems of daily life as well as the ways in which we do scientific research. If, with Karl Popper, we regard all life as a problem-solving activity, the philosophy of science informs all aspects of our existence. It can help us to address the problems of human society as whole—to meet the challenges of developing and implementing new technologies, and even social and political policies. It may profitably be extended to any other fields of knowledge that encompass a problem-solving dimension—to design, the crafts, and even the world of the arts and the humanities.

Chapter 2
It's About Attitude and Experiments

Abstract Although it is clear that philosophers are unable to identify unique methodological elements that characterize scientific research, it appears that we can identify certain features from which it derives its explanatory power. An awareness of the philosophical and historical underpinnings of science provides the researcher with the conceptual and analytical tools to examine and evaluate his practice. These tools will also help in the assessment of the work done in his own field, and in other disciplines.

> You look at science (or at least talk of it) as some sort of demoralising invention of man, something apart from real life, and which must be cautiously guarded and kept separate from everyday existence. But science and everyday life cannot and should not be separated. Science, for me, gives a partial explanation for life. In so far as it goes, it is based on fact, experience and experiment.
>
> Rosalind Franklin (1920–1958).

There can be no serious doubt about the success of modern science in describing and understanding the physical world. The remarkable increase in scientific knowledge since the time of Galileo has been at the root of a technological revolution that has transformed the way we live. It takes only a moment of reflection on the impact of scientifically based improvements in the fields of engineering, medicine, dentistry, transport, and communications to make this clear and undeniable.

The achievements of the scientific program have inevitably stimulated questions about why science has been so successful. The enterprise of scientific research seems to be invested with a special status in which scientists are accorded particular respect because of their access to technological knowledge and the power that derives from it. On the face of it, there would appear to be something special about science that distinguishes it from other forms of inquiry, and much effort and thought has been directed towards identifying and understanding its distinguishing features. The assumption has been that science is characterized by a "method" that allows it to arrive at descriptions of the natural world which have a uniquely powerful explanatory and predictive power.

P. Truran, *Practical Applications of the Philosophy of Science*,
SpringerBriefs in Philosophy, DOI: 10.1007/978-3-319-00452-5_2,
© The Author(s) 2013

For the last four hundred years, philosophers have attempted to elucidate the nature of scientific method. They have tried to construct clear, logical pictures of the way science works to deliver reliable knowledge. However, studies in the history of science and of current scientific practice reveal a complex reality marked by disciplinary peculiarities and pluralistic approaches. Science is, indeed, a messy business. Every attempt to provide a rigorous basis for distinguishing science from non-science, or to provide a description of scientific method which would be embraced by the community of researchers, has ultimately foundered. Indeed, the project to find an acceptable, universal description of scientific method is probably doomed to failure. While a minority of scientists acknowledges a correspondence between the descriptions provided by philosophers with the realities of their own research practice, most researchers pay scant attention to such explanations. However, although the philosophy of science may never succeed in pinning down what scientists actually do, it can provide researchers with more effective and accurate ways of talking about their work. Furthermore, it can tell them what they should do in order to be better scientists. In other words, it can provide researchers with "normative" advice. And that advice will be more pertinent if researchers engage with philosophers in the debate about what they should do and what they actually do.

Not all science is good science. It is easy to point to gross examples such as the ideologically driven application of Lamarck's theory of the inheritance of acquired characteristics, by the Russian pseudoscientist Trofin Lysenko, which contributed to ruinous agricultural practices during the Stalinist era in the Soviet Union. However, poor scientific practice is also at the root of the studies in the 1990's which provided the basis for arguments against the use of childhood vaccinations for whooping cough, the Cold Fusion debacle, or the withdrawal from the market of numerous pharmaceuticals which were shown to have serious, sometimes lethal, side-effects. In a report published in 2005 (Ioannidis 2005) it was estimated that more than 50% of research findings could not be relied upon because of small sample sizes, poor study design, researcher bias, selective reporting and other problems. Although some bad science may be the result of deliberate deception driven by commercial pressure or personal ambition, much of it is rooted in a faulty understanding of the nature of science.

It is evident that researchers who declare that they know nothing of the philosophy of science may, in reality, talk in philosophical terms about what they do. There is always an ongoing debate about the methods of science, about ethical issues related to the use and misuse of science, about its truth status and about issues such as the validity of Intelligent Design, or the causes of global warming. Furthermore, the ways in which scientists articulate and reflect upon their practice will affect, to some degree, what they do in their research. This is an important point. For example, the researcher who describes gathering data before formulating the solution to a problem may not actually be doing this in practice. He is most likely to be making observations in the light of a hypothesis that he has not explicitly formulated or "spelled out". If this is the case, the quality and number of observations may be compromised. The researcher's understanding about what he is doing *as a scientist* may therefore affect the quality of the research being done. The normative prescriptions of the philosophers of science can help the well-intentioned scientist to avoid pitfalls and improve research practice.

Such principles and prescriptions will always have their shortcomings. They may be more appropriate in one research context than another and may fall short in terms of their general applicability. However, they provide a framework, to be continuously and critically reviewed, for the ways in which the scientist may proceed. Generally, it is in the work of those philosophers of science (such as Popper and Kuhn) who have been embraced and recognized by the scientific community that we find an expression of those principles that are likely to be most useful and pertinent to the researcher.

Although philosophers are unable to identify unique methodological elements that characterize scientific research, they can identify certain features from which it derives its explanatory power. These "normative" principles may be summarized as follows:

1. Science is a system of methods for solving problems.
2. Science is characterized by an *attitude* that values intellectual honesty, integrity and open-mindedness, and exhibits a measured skepticism which embraces criticism and rejects dogmatism. It does not accept explanations that make reference to miracles or the supernatural. All scientific knowledge is regarded as provisional. This attitude is not confined to science, and it may be found as a core doctrine within many other disciplines, for example philosophy.
3. Science is rational. It employs agreed methods of reasoning that allow reliable connections to be made between supporting information and the conclusions to be derived from that evidence. Note that good reasoning does not assure the truth of conclusions since these may be based upon supporting information that is false. Although these methods of reasoning are characteristic of science, they are applicable within other areas of knowledge that may not immediately be considered to be scientific.
4. Science uses *experiments*, investigations or studies to test solutions to problems. Experiments allow us make and test predictions about the behavior of the physical world, and to establish consistent, repeatable ways of interacting with it.

Experiments are the practical embodiment of the critical and skeptical attitudes which are at the core of scientific methodology, an attitude which continually asks, "Might my current explanation be mistaken?" and "Is there, perhaps, an alternative explanation?" Experimentation has become the standard instrument in mainstream science for revealing the errors present in our ideas and for comparing and choosing between different solutions to problems (Mayo 1996). It has become the means for establishing the scientific credibility of disciplines as diverse as sociology, economics, anthropology and psychology.

The key and defining characteristic of an experiment is that it should be a genuine test of the solution to a problem. It is crucial that an experimental test should be set up and conducted in such a way that there is a possibility that the results of the experiment could indicate that the proposed solution is wrong or false. There is no point in doing a "test" on an idea that you know will always confirm that idea, no matter the outcome of the test. The researcher therefore, frames the proposed solutions to scientific problems (or hypotheses) so that they are considered to be "falsifiable", although we shall see that the concept of falsifiability has its own problems. Nevertheless, the aim of the good researcher is to establish experiments that are severe tests of the solutions under consideration. The more severe the test which is passed, the more confidence the researcher can have in the proposed solution.

The concept of "falsifiability" was proposed by Popper (1963) as a logical criterion for distinguishing scientific propositions from non-scientific or "metaphysical" ones. Scientific ideas would be subjected to empirical tests in which they would run the risk of falsification and rejection. Popper was particularly impressed by the case of Einstein's Theory of General Relativity which was subjected to a severe and definitive test in the famous experiment carried out by Arthur Eddington in 1919. Eddington and his team demonstrated during a solar eclipse that light from stars that appeared close to the sun was deviated by the effect of the mass of the sun, a novel prediction derived from Einstein's theory. If the light had not deviated by the amount predicted according to the theory of General Relativity then, in principle, Einstein's theory would have been falsified.

Nevertheless, it was recognized by Popper that, if Einstein's theory had not passed the test of Eddington's experiment, this would not in itself have been a decisive blow against Einstein's theory. Philosophers and scientists agree that, even if an idea or solution fails an experimental test, this may not enough to reject it. The solution we have may be the best we have so far, and there may be no viable alternative. There may be doubts about the instrumentation used to carry out the experiment, or the experiment itself may be flawed. This was certainly the case in Eddington's experiment where some of the results obtained were shown to be anomalous because of problems with one of the telescopes used. For many years, there were serious concerns about the validity of Eddington's results. What is clear, however, is that the scientific attitude requires the scientific researcher to employ all the logical and critical faculties at his disposal in order to carry out tests which are as fair and severe as possible, recognizing that, even then, the results may be misleading or plain wrong. All solutions to scientific problems, all scientific knowledge, are in this sense provisional.

Note that, in some research disciplines, it is not possible to conduct experiments to test hypotheses. This is the case in the so-called historical sciences, among which we can name archeology, paleontology, geology, and cosmology. It is simply not possible to manipulate conditions to test ideas about the origin of the universe or the extinction of the dinosaurs. Instead, much of the research effort in the historical sciences is based upon *counterfactual reasoning* which asks the question "What might be *not* be found if a particular event had not occurred?" As such, inquiries are often directed towards the search for a "smoking gun" that will support the acceptance of a hypothesis (but not eliminate competing hypotheses). A good example of a "smoking gun" is the three-centimeter background radiation that was predicted to exist as a result of the Big Bang. Another is the discovery of iridium and shocked quartz in the Cretaceous-Tertiary boundary that provided evidence of the meteor impact that is believed to have caused the extinction of the dinosaurs sixty five million years ago. If an asteroid impact big enough to cause climate change had *not* occurred, then the observed geological evidence would not be found. These differences in the means used for testing hypotheses in the historical sciences do not reduce their credibility or scientific status (Cleland 2001).

If we characterize scientific methodology as comprising the deployment of a scientific attitude together with the rigorous testing of hypotheses, then it has wide applicability outside of mainstream science. The application of scientific methodology is appropriate whenever, we require our interactions with the physical world to be predictable and replicable. This is clearly the prime concern of the physical scientist who is elucidating the laws that describe and explain phenomena in the natural world. However, when we survey the range of research disciplines, it is clear that the applicability of scientific methodology, and especially the testing of hypotheses, is much wider than we might first anticipate.

As in the natural sciences, we see within the humanities the expression of values—intellectual honesty, open-mindedness and non-dogmatic skepticism—which are key features of the scientific attitude. Nevertheless, the humanities are not generally considered part of the scientific endeavor because they generate problems that cannot be resolved by the use of experimental testing. However, within a number of disciplines in the humanities, for example theology and history, there are areas where the application of scientific methodology is clearly appropriate. We would not usually describe the field of biblical exegesis (the understanding and interpretation of biblical texts) as scientific. However, academic studies which treat the bible as a historical document use a number of approaches which may be regarded as scientific, for example the techniques of textual analysis used to identify the authors of the book of Genesis. Similarly, the fundamental questions posed by historians will never be amenable to resolution by scientific research methodology precisely because historical events cannot be repeated under experimental conditions. However, the use of chemical analyses, carbon dating, DNA analysis and MRI scans have all proved to be important tools for the historian, and the application of these techniques may be done within a context which is purely experimental.

This characterization of science as dependent on the rigorous testing of hypotheses allows us to view activities that took place in the pre-scientific age as being within the scientific tradition. Consider the achievements of the Greeks who developed lead sheathing of ships to protect their hulls, air and water pumps, and the truss roof. The Romans developed concrete, built the dome of the Pantheon, still the largest unreinforced concrete dome in the world, and constructed sophisticated systems of aqueducts. These were the result of processes of trial and error that meet all the requirements for being considered scientific. Whenever practical utility is important in the assessment of a conceptual development, the processes are likely to exhibit the features that characterize modern science. There can be no place for dogmatism or a rejection of test results in the area of ship design or in the construction of large stone buildings. When a ship sinks, or a building collapses, the outcome of the test is beyond dispute. The differences between modern science and that which took place in the "pre-scientific" age are primarily ones of rigor and organization, especially the documentation of theories and the experiments used to test them, and the establishment of institutions for discussion, critique and review of scientific ideas and developments. It is also clear that the success of modern science is attributable in large part to the elimination of supernatural explanations from the process of scientific inquiry. The processes of trial and error that underlay much pre-scientific technological development were confounded by the need to satisfy religious traditions or the demands of supernatural forces.

> The whole of science is nothing more than a refinement of everyday thinking.
>
> Albert Einstein (1879–1955).

An awareness of the philosophical and historical underpinnings of science provides the researcher with the conceptual and analytical tools to examine and evaluate his practice. These tools will also help in the assessment of the work done in his own field, and in other disciplines. It is no longer acceptable to reject research in the historical sciences as unscientific. Indeed, we now recognize that the attitudes and methods of the sciences are ubiquitous, and will be found within many disciplines that we would not otherwise consider primarily scientific. The borders between science and non-science are therefore not clearly marked. However, a clear understanding of the distinguishing features of science will help us to make appropriate judgments about the scientific status of research findings and claims. It will also allow us to identify when it is appropriate to apply the methods of the sciences to our problems and when it is not.

Practical Points

- Take time to step back and think about what you are doing as a researcher. A judicious study of many of the ideas and concepts that form the basis of the philosophy of science will improve your reasoning skills and clarify your approach to your research.
- Develop a non-dogmatic skepticism of the ideas and data that are presented to you, but do not let your criticism be unnecessarily destructive. Be fair in your criticism, and supportive and constructive whenever you can. A reputation for being negative will harm your ability to work with others and may deprive you of support when you most need it.
- Make a point of learning about designed experiments and statistics, and the ways they may be used to study those aspects of the world that can be controlled and measured. Experiments are the classical tools of science, and can bring order and clarity into our interactions with the natural world.

- Ensure that your experiments represent genuine tests of your hypotheses. Stipulate clearly the results and outcomes that would constitute evidence against the hypothesis you are testing.
- If we cannot carry out experiments that would involve the manipulation and control of key variables, then we must ensure that we provide reasoned evidence that supports our hypotheses. This will generally be less definitive than the results of a well-designed experiment.
- Scientific thinking is not solely the province of scientists. We can cultivate an attitude that embraces intellectual honesty, open-mindedness and measured skepticism, and a willingness to subject ideas to rigorous testing in all knowledge disciplines and in all walks of life.
- We can make judgments about when it is appropriate to apply scientific thinking and experimental method to our problems, whether these occur in our professional work as researchers or as part of our daily lives. Sometimes, it may simply not be worth the effort—the problem may be just too trivial. Simple trial and error may suffice. At other times the problem may require the application of the full rigor of the scientific methodology. And then there are the times when we need to abandon the analytical perspective of the scientist—to simply look at the full moon on a still cold winter's night and wonder at its beauty. We need to know when to be poets and when to be scientists. A better understanding of science will allow us to make these judgments more clearly.

References

Cleland CE (2001) Historical science, experimental science, and the scientific method. Geology 29:987–990

Ioannidis JPA (2005) Why most published research findings are false. PLoS Med 2(8): e124. doi:10.1371/journal.pmed.0020124   http://www.plosmedicine.org/article/info_3Adoi_2F10.13 71%2Fjournal.pmed.0020124. Accessed 17 Aug 2011

Mayo D (1996) Error and the growth of experimental knowledge. University of Chicago Press, Chicago

Popper KR (1963) Conjectures and refutations. Routledge and Keagan Paul, London, pp 33–39

Chapter 3
What's the Problem?

Abstract In order to make a new contribution to his field of knowledge, the researcher will need to identify a significant problem within his chosen field, a choice that may have profound career implications. Identify a problem which is of an appropriate scope—neither too safe, nor too ambitious—and which is likely to have a solution.

> Science is essentially a problem-solving activity.
>
> Larry Laudan (1986).

How does the researcher choose a problem to work on? It is clear that the choice of a research field will be important. Many researchers will have been drawn to a particular field of knowledge because of a fascination with ideas, or a desire to make discoveries themselves, or simple curiosity. Others will have a vocational motivation, and the successful completion of a research project will be a step along the way to a successful career. However, in order to make a new contribution to that field of knowledge, the researcher will need to identify a significant *problem* within his chosen field, a choice that may have profound career implications. Any scientist of any age who wants to make important discoveries must study important problems (Medawar 1979).

For most this cannot be a process that is done in isolation. Rather, the choice of a significant research problem is done in the context of the researcher's immersion in a scientific community, and for the novice researcher, the graduate student, this is especially the case. Learning how to do scientific research can be regarded as a process of apprenticeship, a process that, in many respects, resembles the ways in which a craft or trade was traditionally taught over the centuries. In those trades where apprenticeships are still formally recognized, the apprentice may go to school to learn about his trade, to learn the theory. An apprentice auto mechanic will learn about the ways in which the internal combustion engine works, the application of the computerized diagnostic systems that are now used in the auto

P. Truran, *Practical Applications of the Philosophy of Science*,
SpringerBriefs in Philosophy, DOI: 10.1007/978-3-319-00452-5_3,
© The Author(s) 2013

industry, and the use of appropriate equipment to do repairs. However, a key element of his apprenticeship will be learning the thought processes required to solve problems and the use of equipment "on the job". This is the part of the trade that is not taught in the textbooks or manuals, but which is learned from those who have already achieved mastery, those who will teach the "feel" that is required to ensure that a nut is tightened just right (not too slack and not too tight). They will teach the tricks of the trade that ensure that a job goes smoothly, and show the young apprentice how to solve the problems that occur. These skills will be taught by a more experienced "master" mechanic who will also inculcate the correct attitudes required to resolve engineering problems, attitudes that involve a disciplined approach, but which will also require the development of the creativity necessary to resolve more demanding or novel problems.

Now consider the process by which the student learns how to do scientific research. The apprentice researcher will have spent many years learning the knowledge content of his or her field, be it medicine, physics, biology, engineering, psychology or design. An understanding of established concepts, laws and theories will have been deepened by problem solving exercises, and by practical laboratory exercises or investigative projects. This process of learning by "finger exercises" and by doing continues throughout the process of professional initiation, with the student being assigned problems that, in the words of the philosopher Thomas Kuhn, become "progressively more complex and less completely precedented "(Kuhn 1996)". However, it is as a graduate student that the apprentice researcher is required to undertake a project that will be intended to make a significant contribution to the established body of knowledge that represents the epistemological content of his chosen field. *The student is required to find out something new.*

The early exercises carried out in high school or as an undergraduate will have established the foundations for many of the basic skills required for research, and may have provided an opportunity for the student to carry out some creative problem solving. However, in order to learn how to do research as a professional, the apprentice will need to establish a relationship with a master researcher from whom he or she will learn those elements of research practice that are *not* made explicit in the text books. The role of master researcher is taken by an advisor (or supervisor) who, by guiding, pointing and cajoling, will initiate the apprentice into the skills and attitudes necessary for successful scientific research. The advisor will open the way for the apprentice to engage with seasoned researchers and the whole community of scientists who provide the context for the debate and criticism that is the lifeblood of the research process.

For the young apprentice researcher the choice of an advisor will be crucial. It may be difficult to balance the requirement to choose an advisor who provides effective mentoring, with the desire to work in a department where significant research is being done in his field of interest. It will always be difficult to assess whether the relationship with the proposed advisor is likely to work. A brilliant researcher may not be a sympathetic and effective advisor. It is important to avoid choosing someone who presents a project on a plate, a project that may be successfully completed and that may lead to the student obtaining his degree, but one in which the apprentice has been treated more like a technician rather than a trainee researcher. The critical role of the advisor is evident from early on in the research project, when he helps the apprentice researcher to select a problem that is in an appropriate area, and that is neither too safe nor too ambitious.

> Never tackle a problem of which you can be pretty sure that (now or in the near future) it will be tackled by others who are, in relation to that problem, at least as competent and well-equipped as you.
>
> Edsger Wybe Dijkstra (1930–2002)

Where does one look for a significant problem to work on? The researcher will have an interest in a particular field and may be curious about a specific area. However, he will need to deepen his knowledge of the field, and especially his awareness of current research. The researcher is required know the areas in his field of knowledge into which research efforts are being directed and where major shifts are taking place. These will be the areas where new and significant problems will be found.

Furthermore, the researcher will need to identify those areas where significant problems within the field *have already been resolved*. At first, this may appear to be paradoxical, but the resolution of a significant research problem will frequently be the source of a whole new family of problems to be addressed. For example, the use of X-ray crystallography in the study of the protein Myoglobin provided an understanding of the three dimensional structure of proteins that, apart from its application to the study of other proteins, gave rise to a new set of problems concerning the ways in which enzymes interact with their substrates. Similarly, the explanation for the redshift seen in the light from distant galaxies (the universe is expanding) gave rise to new problems concerning the size and age of the universe and whether it will continue to expand forever or ultimately contract in on itself. This proliferation of problems is the reason why the growth of scientific knowledge has been so phenomenal. The solutions to some problems will have major or revolutionary impacts upon the fields of knowledge in which they take place. The identification of the structure of DNA opened up the field of molecular biology to a vast array of new problems including the mechanisms by which replication of DNA takes place, and the ways in which the information locked in the linear sequences of bases in the DNA molecule is used to build proteins. Other waves of problem solving activities were generated by the solutions provided by Newton's mechanics, by Einstein's General Theory of Relativity, and by Darwin's theory of Evolution by Natural Selection.

Such major problem solving achievements may radically change the way in which we see the world. Einstein's General Theory of Relativity introduced geometrical concepts of space-time that eliminated the Newtonian requirement to explain the motion of bodies in terms of gravitational forces acting at a distance. The elucidation of the structure of DNA shifted the focus of biological research towards explanations based upon the molecular structures of living organisms.

Thomas Kuhn described the effects of such major problem solving achievements as *paradigm shifts*, scientific revolutions that fundamentally change the theories, concepts and methodological tools with which researchers in a particular field approach their work. It is Kuhn's view that, once such a paradigm shift has been made, researchers will tend not to question the fundamentals, but will "mop-up" the problems that a paradigm shift leaves to be done, a process that Kuhn describes as "normal science" and "puzzle solving". Kuhn makes the point that "mopping-up operations are what engage most scientists throughout their careers" and that "(F)ew people who are not actually practitioners of a mature science realize…quite how fascinating such work can prove in the execution" (Ibid. 24).

This does give the impression that once the giants of science have spoken, the task left to ordinary scientists is less creative busy work, and Kuhn's view of seismic shifts in scientific knowledge has been widely criticized. It has been argued (Toulmin 1969) that revisions in scientific knowledge take place much more frequently and less dramatically than in Kuhn's model of revolutionary and normal science, and that they take place during periods of what Kuhn would call "normal science". Furthermore, the problems that are exposed as a result of a major paradigm shift may themselves be of paramount significance. Kuhn's rather disparaging reference to the "mopping-up operations" that engage most scientists throughout their careers would include, in the case of molecular biology, the groundbreaking research that revealed the existence and functions of the molecules and structures involved in the assembly of proteins, or the work done to elaborate the mechanism by which DNA is replicated during division of cells. These seminal achievements can scarcely be described as "mopping up" and they will themselves have generated substantial programs of challenging and significant problems.

The reality for most apprentice researchers is that they will be doing work that is part of a large research program, where individual researchers will typically be working on related problems and using similar methodologies. The apprentice will most likely choose a significant problem that will fit into the ways of doing research of the department. It will be done within the current research paradigm and will utilize the experimental methods and facilities available. The apprentice researcher will have to find a balance between working closely within the program of inquiry in which his department of choice is engaged, and establishing where the boundaries of knowledge lie for his field and positioning himself there. His advisor should provide help and guidance through the minefield of the literature, and help to identify a problem with the appropriate scope and feasibility.

It may seem prudent to keep one's head down and focus on the project in hand. However, there will be a real benefit to the researcher in developing an understanding of the broader problem context, the paradigm within which the work is being done. The attempt to engage with the problems that underlie the overall research program will furnish the apprentice researcher with the opportunity for dialogue with his advisor and with other researchers. It will provide motivation during the difficult days ahead and will engender a wider, more critical perspective. This may allow the development of ideas that will make a significant impact within the field. The advice of Sir Karl Popper is highly pertinent: "Try to learn what people are discussing nowadays in science. Find out where the difficulties arise, and take an interest in disagreements. These are the questions that you should take up. In other words, you should study the *problem situation* of the day. This means that you pick up, and try to continue, a line of inquiry which has the whole background of the earlier development of science behind it" (Popper 1974).

Remember that many of the most important discoveries are made by young researchers who bring a fresh, uncluttered perspective to their work: Albert Einstein was in his mid twenties when he published his work on the Special Theory of Relativity; Alfred Sturtevant, who discovered the principle of genetic mapping, was in his early twenties; Charles Darwin was in his early twenties when he formulated the ideas of natural selection that culminated many years later in the publication of On the Origin of Species. Such ambitions, however, also have their dangers. There may be temptations to explore radically new ideas or embark on a project that has too ambitious a scope. In the twenty-first century, with the scale and profusion of research in all fields, the guidance of a sympathetic and experienced advisor will be of critical importance.

Practical Points

- You will probably want to make an impact by contributing to the resolution of a significant problem. However, it is important that you identify a problem that is of an appropriate scope—neither too safe, nor too ambitious—and that is likely to have a solution.
- Identify an area of knowledge that is within your field, and that is within the area of interest of the research group or department in which you are planning to work. There is no point in choosing a problem that is outside the range of the research interests of your advisor and the group of researchers with whom you are going to work and from whom you will be seeking assistance.
- Recognize that, as a graduate student, you are in a process of apprenticeship. Choose an advisor with whom you will be able to work effectively and who will teach you the practicalities of research. Learn as much as you can from those around you with hands-on experience, your advisor, fellow students, and (especially) the expert technicians that you may work with.
- When you do your literature search look especially carefully at those areas where significant problems have recently been solved. These are places where there may be a proliferation of new, exciting and fruitful problems.
- Make yourself aware of the experimental systems available within the research institution in which the work will be done. It will be more difficult for a researcher to work outside of the experimental models and instruments that are available within the research group of your choice.

References

Kuhn T (1996) The structure of scientific revolutions. University of Chicago Press, Chicago, p 47
    (It could be argued that this is essential reading for any researcher. In this highly readable book
    Kuhn introduces the concept of the scientific paradigm that has been so influential within and
    outside the world of science. Chapters II and III provide clear introductions to the ways para-
    digms may change, and to the nature of "normal" science.)
Laudan L (1986) Progress and its problems. University of California Press, Berkeley, p 11
Medawar P (1979) Advice to a young scientist. Alfred P. Sloan Foundation Series, New York
Popper KR (1974) Conjectures and refutations. Routledge and Kegan Paul, New York, p 129
    (Popper's writing is refreshingly lucid, but demands a willingness to accommodate references
    to the whole of the Western philosophical tradition. Nevertheless, most of the essays in this
    collection are highly accessible to the general reader.)
Toulmin S (1969) The uses of argument. Cambridge University Press, Cambridge

Chapter 4
Oh No! Something Doesn't Fit!

Abstract Establishing an appropriate attitude towards anomalies and deviations provides a powerful basis for identifying new problems and for the effective characterization of research problems. Negative or unexpected results may attain a significance that is as great as positive results because they provide the opportunity for the development of new, or even paradigm-shifting, hypotheses.

> Anomaly appears only against the background provided by the paradigm. The more precise and far-reaching that paradigm is, the more sensitive an indicator it provides of anomaly and hence of an occasion for paradigm change.
>
> Thomas Kuhn (1922—1996).

Researchers work within an accepted framework of theories, concepts, and problem-solving methods, a *paradigm*, which defines the field in which they are doing research (Kuhn 1996). For example, the current paradigm within which cosmologists work includes an acceptance of the evidence for an expanding universe that began with the Big Bang, the validity of the use of optical and other telescopes for making observations of electromagnetic radiation, and the application of the theory of relativity in order to understand the events that occurred in the early days of the universe. The Big Bang model has been successfully used to explain the mutual recession of galaxies, and it predicted the cosmic background radiation that remains as evidence of this cataclysmic event. It has been confidently applied to the estimation of the age of the universe and the detailed description of events that occurred within the first seconds and microseconds after its birth.

Typically, as work progresses within a field of research, problems and observations may emerge that threaten the paradigm defining the knowledge structure of the field. In the case of the Big Bang theory, astronomers showed, in the 1970s and 1980s, that there was insufficient visible matter in the universe to account for the apparent strength of gravitational forces within and between galaxies. Furthermore, the universe today is far more lumpy and contains far less deuterium

P. Truran, *Practical Applications of the Philosophy of Science*,
SpringerBriefs in Philosophy, DOI: 10.1007/978-3-319-00452-5_4,
© The Author(s) 2013

than can be accounted for by the Big Bang model. However, these anomalies did not lead to an abandonment of the Big Bang paradigm. Rather, cosmologists proposed the idea that up to 90% of the matter in the universe is composed of a material (dark matter) that does not emit light, and cannot be detected by instruments that collect electromagnetic radiation. While dark matter was initially controversial, and as yet there is no direct observational evidence for its existence, it is now indicated by numerous observations such as the unevenness of the cosmic microwave background radiation, large-scale structure distributions in the universe and X-ray measurements of galaxy clusters. Although there is now convincing circumstantial evidence for the existence of dark matter, the jury remains out.

It may be tempting to explain away anomalies, or to modify currently accepted theories in order to accommodate the new observations in an "ad hoc" fashion. Auxiliary hypotheses may be "bolted on" to the troubled theory. The development of the concept of dark matter may therefore be regarded as an auxiliary hypothesis within the Big Bang theory. Such auxiliary hypotheses may gradually accumulate until the original theoretical framework begins to buckle under the weight of all the exceptions and *caveats* with which it is burdened. As the anomalies accumulate they point towards the opportunities for significant developments that may challenge, or even supplant, the current paradigm within which researchers in the field are working.

In "The Structure of Scientific Revolutions", Kuhn describes how the progressive accumulation of anomalies within the Ptolemaic model of the universe paved the way for the Copernican Revolution. Ptolemy's description of the solar system in the second century BCE placed the earth at the center with the Sun and the planets in circular orbits around it. Because the planets move at different velocities in their journeys around the Sun, it appears from the Earth that they sometimes reverse their motion relative to the background stars in the sky, a phenomenon called "retrograde motion". Ptolemy devised an ingenious system of "epicycles" (circles upon circles) in order to explain retrograde motion. Nevertheless, as observational data accumulated over the centuries, astronomers found increasing numbers of anomalies—discrepancies between the orbits predicted by the Ptolemaic model and their own observations. These indicated that something was badly wrong with the Ptolemaic system, and they paved the way for a radical shift in perspective that placed the sun, rather than the earth, at the center of the solar system (the Copernican system) in the sixteenth century.

Isaac Newton's mechanics provided a description of the physical universe that provided stunningly powerful methods for the calculation of the orbits of the planets, the trajectory of missiles and the forces acting upon mechanical structures. It even allowed the observational confirmation of the existence of a hitherto unsuspected planet, Neptune, which was predicted as a result of perturbations seen in the orbit of Uranus. The intellectual world of Enlightenment Europe was in thrall to the explanatory power of Newton's view of the physical world. However, it emerged that not all was right. There were a number of problems that could not be resolved. Among them was the observation that the precession in the perihelion of the planet Mercury, as it orbited the Sun, could not be fully explained by

Newton's theories. Attempts were made to explain away the (tiny) discrepancy. It was proposed that there might be dust between Mercury and the Sun, or a hitherto undiscovered planet, speculatively named Vulcan, orbiting the Sun inside the orbit of Mercury. It took the development of Einstein's General Theory of Relativity to accurately account for the observed precession. It is now recognized that, despite the fact that Newton's mechanics provides an effective model for day-to-day calculations of the way bodies move, being used even for the calculations required to send spacecraft across the Solar System, it is Einstein's theory that provides the more accurate and ultimately more truthful account of the way bodies move in the universe. It is the potential fate of many theories, even those, like Newton's, with great explanatory power, to be superseded.

In discussions of the nature of science, the term "anomaly" is used to indicate a phenomenon that cannot be accounted for within the current paradigm and that might provide a significant threat to it. The term "deviation", on the other hand, is used to refer to a phenomenon or observation that is unexpected but which will not pose any real challenge to the validity of the current paradigm. It is a term widely used within the engineering and manufacturing industries to identify performance that is outside of specifications. For example, an increased number of quality control failures in testing at the end of production run will constitute a deviation; so will a batch of product that generates a higher than expected number of customer complaints. The identification of the deviation will initiate a program to find the cause and implement corrective action. It will not challenge the scientific or engineering principles upon which the manufacturing system is built.

The concept can also be usefully applied to research. An unexpected experimental result would constitute a deviation. The most likely explanation would be that a mistake had been made in the experimental procedure, and the immediate action would be to ensure that the experiment had been correctly performed. However, if the result were to be confirmed it might constitute an anomaly. For example, the discovery of pulsars by Hewish and Bell in 1967 provides a good example of a deviation that turned out to be a significant anomaly. Jocelyn Bell identified unusual, rapid periodicities in radio telescope signals. She established that the pulsing radio source was *not* a deviation that could be attributed to man-made interference or errors in her equipment. It originated from outside the solar system. However, these rapid pulses, 1.3 seconds apart, appeared to be incompatible with a rotating object as large as a star and therefore constituted an anomaly. The objects (now called pulsars) were explained as being produced by much smaller, denser, rapidly rotating neutron stars. This was an important anomaly that opened up new areas of inquiry, but it was not one that threatened the then current paradigm of stellar evolution.

Anomalies and deviations provide us with an opportunity for identifying new problems. The researcher looks out for anomalies, for deviations from what is expected. Instead of trying to explain these deviations away by ad hoc hypotheses, such as the dust proposed to account for the discrepancies in the orbit of Mercury, he sees that such anomalies represent an opportunity for discovery. The recognition of the importance of anomalies provides an incentive to design robust experiments that will clearly confirm their genuineness. This means that negative or unexpected results may attain a significance that is as great as positive results because they provide the opportunity for the development of new, or even paradigm-shifting, hypotheses.

Consider the celebrated work of John Snow (Vinten-Johansen et al. 2003) who in 1849 identified that cholera was a disease spread by contaminated water. In Snow's day most physicians believed that cholera was caused by "miasmas", poisonous gases that were thought to arise from sewers, swamps, garbage pits, open graves, and other foul-smelling sites of organic decay. According to this account cholera would be expected to occur in those areas and households that were the most unsanitary. Snow observed anomalies in this explanation when studying the cholera outbreak in London in 1853. In the area affected by the cholera outbreak, there were only a few deaths at a workhouse (an institution offering work and employment to the poor) compared to the death rate in nearby houses. Those favoring the miasma theory had expected a higher than average rate at the workhouse since the inmates were poorly nourished, unclean, and of "low morals", indicators, according to the miasma theory, of susceptibility to disease. Snow realized that these anomalies cast serious doubt on the miasma theory, and they pointed him towards the "germ theory" that held that diseases might be spread by invisible organisms. Snow was familiar with this alternative theory that was generally unpopular at that time. During the outbreaks of cholera in London in 1848 and 1853, he carried out the pioneering epidemiological studies that identified the mechanisms by which cholera was spread in the water supply.

Snow's ultimate success was arguably due to his being prepared to respond appropriately to the anomalies that he identified.

1. He clearly had a sound grasp of his subject. He understood the miasma theory of cholera and was well acquainted with the controversial germ theory. His understanding of the miasma theory, and its predictions for the distribution of the disease, allowed him to recognize the anomalies in the pattern of the disease among the inmates of the workhouse in London. If he had been less acquainted with the prevailing theory and its implications, the deviations may not have come to his attention.
2. Snow took the anomalies seriously and regarded them as pointing towards a problem to be solved. He did not attempt to explain them away, but recognized the possibility that the prevailing orthodoxy was flawed and that the germ theory might provide a better explanation for the spread of the disease.

It may not always be obvious that that the generation of significant research problems derives from the recognition of anomalies or deviations. For example, when the astronomer turns his telescope to a particular part of the night sky, it may not seem as if he is doing so because of a recognizable anomaly. He may be cataloguing deep space objects to provide material for an improved stellar atlas. Such systematic astronomical surveys are done as a response to known theoretical and observational problems that cannot be resolved without a more accurate catalogue of objects in the universe. However, it is more likely that he will be participating in a program established to resolve a problem generated by anomalies. Examples of such anomalies may be deviations from the predicted distribution of matter based on current theories of the origin of the universe, or the problems associated with the discovery of objects exhibiting strange or unexpected behavior. The motivation for such systematic gathering of data will ultimately found within the problems, anomalies and deviations that drive the research efforts in the field of astronomy.

Similarly, a botanist may be carrying out a research study that involves making random collections from a habitat in order to provide evidence that there are plants present that will justify a protected status for the site. It first sight there would appear to be no immediately discernable deviation or anomaly that would be the basis for the study. However, the research may likely be rooted in the recognition of something unexpected, for example the discovery of a rare species during a preliminary survey of the site. Such a systematic botanical survey may be part of a wider, statewide series of surveys aimed at identifying habitats that need protection. In this case the research study may be a response to a wider problem, for example, the destruction of habitat by recreational all-terrain vehicles in areas that had hitherto been considered too remote to be endangered. The evidence of environmental damage represents a deviation from theoretical assessments and predictions that will have provided the basis for state environmental policy. It will be essential for the researcher to be aware of the studies on which state policy was based, and the nature of reported environmental injuries (deviations) that have occurred. This will allow him to conduct the most effective research to derive new environmental models and provide more powerful predictors of environmental stability.

In the world outside of mainstream science, the recognition of anomalies or deviations frequently provides the starting point for the identification of problems. When an automobile refuses to start in the morning, we have a deviation from expectations—a problem. On a manufacturing line, the identification of product which deviates from the acceptable range of specifications again represents a deviation from expectations and a problem. The housing researcher making a survey of housing provision in immigrant neighborhoods may well be acting in response to problems of localized crime, poor health and low educational performance that represent deviations from the expectations of policy makers and their advisors.

Establishing an appropriate attitude towards anomalies and deviations provides a powerful basis for the effective characterization of research problems. At a prosaic level, the answer to the simple question "What has changed from what we expected?" provides a clear focus for problem resolution. On the other hand, the identification of significant anomalies helps the researcher select worthwhile, important problems that may ultimately result in a major paradigm shift within his chosen field. It provides an incentive for the researcher to understand where the boundaries of knowledge lie, to identify the unsolved problems, and to do research that is truly pushing those boundaries into new territory.

Practical Points

- Deviations and anomalies may provide a powerful clue as to where your next research problems will lie. The identification of deviations from what is expected also provides a starting point for the resolution of problems.
- Deviations or anomalies may be due to errors in your thinking, mistakes in your methodology or interferences by hitherto unsuspected factors. Eliminate other possibilities before declaring that you have identified an anomaly. This can best be done by careful experimental design and technique.
- Do not ignore, or cover up or work around anomalies. You may be missing an opportunity to discover something new.
- Be aware of the paradigm in which you are operating. The detection of anomalies that challenge the current paradigm in your field may provide the opportunity for novel and significant research.
- Do research that allows the possibility of well-established negative results. These will be more useful than confirmations that do not allow the possibility of the detection of anomalies. Solid experimentation allows the detection of errors in currently accepted theory.

References

Kuhn T (1996) The structure of scientific revolutions. University of Chicago Press, Chicago
Vinten-Johansen P, Brody H, Paneth N, Rachman S, Rip M (2003) Cholera, chloroform, and the science of medicine: a life of John Snow. Oxford University Press, Oxford

Chapter 5
The Solutions to Problems

Abstract Hypotheses are tentative solutions to problems. They are guesses, or conjectures, but they are not wild guesses; they are guesses made by informed and creative minds. Even after such rigorous testing any scientific theory or hypothesis may be ultimately replaced one that is superior. In this sense all scientific knowledge is regarded as being provisional.

> A hypothesis is a statement whose truth is temporarily assumed, whose meaning is beyond all doubt.
>
> Albert Einstein (1879–1955).

The researcher who has defined an interesting, worthwhile problem will try to identify solutions to it. These proposed solutions are called *hypotheses*—and they are called still called hypotheses even if they have passed the most rigorous testing. Isaac Newton's system of mechanics, his marvelous solution to the problem of explaining the motion of bodies in space, is a hypothesis. Mendel's law of independent segregation of inherited characteristics is a hypothesis, as are Quantum Mechanics, and Wegener's Theory of Continental Drift. At a less lofty level, statements such as "This testing system provides improved sensitivity and specificity when compared with currently approved systems", "The use of the color green in decorative schemes is soothing and relaxing" and "Immigrants experience prejudice when applying for accommodation" are all hypotheses. Some hypotheses may be called *theories*, and sometimes these two terms are used interchangeably. However, generally speaking, the term "theory" is reserved for hypotheses of wider explanatory scope such as Darwin's Theory of Natural Selection or John Maynard Keynes' Theory of Economics. The word is also reserved for hypotheses that have become generally accepted by the scientific community.

It may seem strange that we use the words "hypotheses" or "theories" to describe solutions to scientific problems that have become so widely accepted and applied. It might be considered that theories such as Darwin's or Newton's deserve

P. Truran, *Practical Applications of the Philosophy of Science*,
SpringerBriefs in Philosophy, DOI: 10.1007/978-3-319-00452-5_5,
© The Author(s) 2013

to be regarded as facts or laws. However, the terms hypothesis and theory, rather than fact or law, convey an attitude to the status of scientific knowledge that is entirely appropriate. All scientific hypotheses are considered to be tentative until they have been submitted to the rigors of empirical testing. Even after such rigorous testing and corroboration it is recognized that any scientific theory or hypothesis may be ultimately replaced by a superior hypothesis. In this sense all scientific knowledge is regarded as being provisional. This may be at odds with the more general, public understanding of science, where an explanation of a phenomenon (for example, evolution by natural selection or the greenhouse effect on global warming) may be condemned as being "just a theory" or "just a hypothesis and not a fact". However, in the scientific world *all* explanations are hypotheses or theories. It might be thought that the term "fact" should be confined to simple observations, but as we will see later, even such statements as "The sun set at 6.30 pm" have hypothetical or theoretical underpinnings.

> The simplest and most obvious cause which can there be assigned for any phenomena, is probably the true one.
>
>   David Hume, English philosopher (1711–1776).

Once the researcher starts the process of trying to identify the solution to a particular problem, there are likely to be many hypotheses that appear to be possible solutions. It is the task of the effective researcher to generate hypotheses - and then to rigorously evaluate them prior to any experimental testing. This process of weeding out hypotheses will first of all require the consideration of their plausibility, and internal consistency. A simple rule for selecting the most plausible hypothesis is "Occam's Razor" that says that the simplest explanation is likely to be the best. The most plausible hypothesis is also likely to be the most "conservative" or "parsimonious", meaning that it will not introduce new concepts when old concepts will do. Hypotheses should also exhibit internal consistency; they should be framed clearly. The effectiveness of the initial assessment of hypotheses using these criteria will be determined by the researcher's experience and judgment. The more familiar he is with the content of his field, the easier it will be to identify the most promising hypotheses. It is arguable that what distinguishes great researchers is not just the ability to generate powerful hypotheses, but that they possess the ability to rapidly identify those that are not worth pursuing and then to eliminate them. Lesser minds will cling jealously to the objects of their creation, and will not subject them to the critical assessments that they deserve.

Within the realm of possible hypotheses, the researcher will be trying to identify those that are important. A significant hypothesis should exhibit good explanatory power; it should predict something unexpected or difficult to predict. Consider the widely accepted theory that it is possible to locate water underground by use of a divining rod. Water divining has been shown to have a 70% chance of success, which appears to be pretty impressive. However, other studies have demonstrated that if sites are simply chosen at random, water is still found with a 70%

success rate. The explanatory power of the theory of water divining is therefore zero. On the other hand, people without a good knowledge of astronomy will find it difficult to predict the time of the next solar eclipse. The probability that they might predict that an eclipse will occur at any particular time is very low indeed. However, the use of astronomical theory will allow such predictions to be made with a high level of confidence, attesting to the high explanatory power of the Newtonian mechanics used to make the calculations (Rapoport 1972).

It is also worth noting that hypotheses with a high explanatory power will also yield novel predictions that take risks. In the case of our previous example, a theory based on astronomical ignorance may make the prediction "A solar eclipse will occur sometime at some point on the earth in the next 10 years", a prediction that has a high probability of success, takes no risks and provides very little useful information. On the other hand, the prediction that a total solar eclipse will take place on Monday, August 21, 2017 and be visible in Christian County, Kentucky, is one that puts itself on the line and has a low degree of probability. It may seem paradoxical, but it is a characteristic of hypotheses with high explanatory power that that they make predictions that are daring, informative and highly unlikely, that is, they would be unlikely to be guessed at before the development of the hypothesis. Thus, one of the deductions from Einstein's Theory of Relativity was a novel prediction that light passing close to the sun would be deflected from its path by an amount significantly greater than that predicted by Newton's theory. This prediction was bold and, in terms of Newton's theory, improbable. In 1919 Arthur Eddington made observations of deflections in the apparent positions of stars as their light passed close to the sun at the time of an eclipse. The observed deflections were in line with Einstein's predictions and provided a powerful corroboration of it.

One of the key features that seem to mark a hypothesis as scientific is that predictions or deductions derived from it are empirically testable or subject to potential falsification. If a test is to be valid there must be possible circumstances under which the subject of the test might fail it. A test procedure that always gives a positive result cannot be taken seriously as a test. As we have seen, Einstein's prediction of the deflection of light passing close to the sun provided a powerful test of the Theory of Relativity. In principle, if the deflection had not been observed, a key prediction of Einstein's theory would have been falsified, and the validity of Relativity would have been cast into doubt. On the other hand, any failures of predictions derived from the analysis of astrological charts do not threaten the edifice of astrology. They are more likely to be explained away by uncertainties in the data relating to the subject's time of birth or by an appeal to the potential complexity of the interactions of all the celestial forces involved.

Although the principle of falsifiability provides a useful guideline for research, in practice things are often more complicated. The inability of Newtonian mechanics to adequately explain the orbit of Mercury did not lead to its rejection. Newton's system was far too successful to fall as a result of the identification of this single anomaly and, at the time, there was no alternative to Newtonian physics that matched its explanatory power. In a similar fashion, the measurements of the deflection of light that Eddington obtained in the test of Einstein's theory were by no means unequivocal. One set of Eddington's observations that happened to be in line with the deflection predicted from Newton's theory was rejected on technical grounds. An explanation was advanced that the telescope used for these observations had suffered distortion due to the effect of the sun. This *auxiliary hypothesis* was invoked in order "save the phenomenon". After protracted debate (including alternative explanations for the observed deflections) it was accepted that the results provided experimental confirmation of Einstein's theory. Nevertheless, even if the results had *not* confirmed the prediction, it is unlikely that the theory would have been rejected outright. Einstein's explanation of the perihelion of Mercury also provided powerful corroboration, and there was a beauty and persuasiveness about the theory that carried all before it.

The requirement that hypotheses be falsifiable is not a hard and fast rule, but it does provide an underlying procedural logic for the scientific method - we need to be rigorously critical and skeptical of our hypotheses. We need to frame them so that they are testable, and to subject them to severe challenges that include the possibility of their falsification. A simple, unsophisticated view would be that if experimental evidence refutes a hypothesis, then it should be rejected. However, it is widely understood that this view is, in reality, untenable since the so-called "facts" or observations that are used to refute hypotheses are themselves theory laden. There are no hard facts! Lakatos has given an imaginary example of observations made with a radio telescope being used to test a particular theory of gravity (Lakatos and Musgrave 1970). If the observations were inconsistent with the theory, we would consider it to be "falsified", but we would not necessarily challenge the complex theory of radio-optics on which the observations depend. However, observational or experimental data are themselves fallible, since they also involve fallible theories. Nevertheless, they are generally taken as unproblematic background knowledge in the testing and selection of theories.

A more sophisticated view of falsification recognizes that science is both critical and fallible. This perspective recognizes that it takes more than a single experiment or test to falsify a hypothesis. The falsification of a theory or a hypothesis may not be enough to warrant its rejection. It may fail in certain respects, but it may be better than the alternatives, if there are any. Thus we do not declare a theory to be falsified before the emergence of a better theory. Rather, research involves making decisions about choosing between available hypotheses, with the researcher making the tests as severe as possible. In a similar fashion, a hypothesis will not be *accepted* as the result of its corroboration by a single confirmatory experimental test. Rather, our confidence in a particular hypothesis or theory will be based upon a network of inter-meshing theories and observations, each of which will be held as being worthy of our acceptance.

Finally, where do hypotheses come from? How do researchers come up with the new ideas that will make a significant contribution to, or even revolutionize, their field of knowledge. It is evident that the evaluation of a newly generated hypothesis will involve logically-based procedures for its testing and corroboration (*a logic of justification*) which includes the deduction of testable predictions, the assessment of internal consistency and the evaluation of supporting data (Achinstein 1977). However, there appears to be no established logical mechanism for the generation of hypotheses, no *logic of discovery*. In the 17th and 18th centuries, philosophers (Bacon, Boyle, Locke, and Newton) tried to formulate rules that would lead to the discovery of facts and theories about nature. These rules centered on the belief that hypotheses could be derived from observations, usually by a process of *induction*. The gathering of unbiased observations would allow the scientist to produce generalizations and laws that revealed the truth about nature.

Although this project was, by and large, abandoned by philosophers by the end of the 19th century (Laudan 1981), the development of powerful, pattern-seeking algorithms has revived the debate about the feasibility of inductively generating testable hypotheses. This has recently been re-asserted in the context of the human genome project, where the generation of vast quantities of data is seen as the starting point for the computerized analyses that it is hoped will ultimately provide an understanding of human biology (Allen 2001). These data provide a reservoir of observations that will be mined to yield the patterns and relationships that will eventually be generalized into hypotheses and theories. These programs are certainly able to identify general patterns in data, but it may be argued that such procedures are actually testing the applicability of the hypothetical assumptions enshrined within them. Induction itself is incapable of providing a logical route to useful scientific hypotheses. On the contrary, the emphasis on the making of observations as a starting point for the generation of scientific knowledge is widely regarded as *practically* misconceived. This is neither the way in which researchers actually do work nor should work.

The process by which hypotheses are generated is best investigated as a problem of psychology rather than logical analysis. Hypotheses may come as flashes of inspiration, as dreams or as the result of lateral thinking and the linking of disparate concepts. There are anecdotes from the history of science that give some weight to this view. There is the famous dream in 1865 of the chemist Friedrich Kekulé who reported that he conceived the solution to the structure of the organic chemical benzene after an intense period of pondering the problem. While dozing in front of the fire, he saw a vision of a snake eating its own tail. The circular form suggested the solution to his problem, the ring structure of benzene. Other scientists have also made discoveries as a result of dreams. The Russian scientist Mendeleev had a dream that suggested that the elements could be organized by atomic weight to create the Periodic Table. Otto Loewi dreamed the design of his elegant and ingenious experiment to demonstrate the effects of acetylcholine and adrenaline in respectively slowing down and stimulating the heartbeat in a frog. Loewi jotted down his dream but couldn't decipher his own notes. The following night the dream returned and he was able to write notes that gave him the design he was seeking.

The interactions between researchers can be particularly productive. Watson and Crick arrived at the double helix structure of DNA by a process of physical model building, heated discussion and the interpretation of clues provide by other scientists. The example of Maria Mayer, who shared the Nobel Prize for physics in 1963, is especially instructive. Mayer was working on the problem posed by the recognition that certain "magic" numbers of protons and neutrons in the atom were capable of conferring stability on an element by making up a complete shell within the atomic nucleus. Although Mayer was working as a senior physicist in the Theoretical Physics Division at the University of Chicago, her background was actually in the field of chemical physics, much of her previous work having been in the separation of isotopes. Her knowledge of nuclear physics was acquired by discussion with colleagues rather than bookwork. This allowed her to have a freshness in her approach that was not constrained by traditional approaches to nuclear theory. Crucially, Mayer was fully acquainted with the experimental facts and had been immersed in this problem for a number of years. The key moment came when she was casually asked a question about the problem by Enrico Fermi, who then left the room at answer a telephone call. By the time he returned, 15 min later, Mayer had experienced a flash of intuition that solved the problem of "magic" numbers that had proved to be so intractable to her colleagues.

It is generally agreed that the generation of hypotheses is a creative process. It takes knowledge, experience, skill, and intuition to come up with a great hypothesis, just as it takes knowledge, experience, skill, and intuition to paint a masterpiece picture or compose a symphony. Hypotheses are guesses, or conjectures, but they are not wild guesses; they are guesses made by informed and creative minds. Like the artist, the researcher needs to provide an environment for creativity, the quiet reflective state in which ideas may be generated, or the conversations and discussions in which sparks may be generated by the engagement of minds. This all requires commitment, energy, perhaps even obsession, and when the ideas come, a willingness to evaluate them ruthlessly!

Practical Points

- Hypotheses are tentative solutions to problems, and any particular problem may have a large number of possible solutions. Create the best conditions for the generation of hypotheses—study, think about your problems and discuss them with others.
- Allow yourself periods of relaxation, semi-dreaming so that the ideas will flow. The more hypotheses that are generated, the more likely it is that you will discover those that will take you closer to the truth.
- Frame your hypotheses as clearly and unambiguously as possible.
- Make sure you have a comprehensive knowledge of your field. Challenge your new hypothesis by assessing how it stands up in the face of currently accepted hypotheses and observations. If your hypothesis is contradicted by other facts or observations, consider if such observations may have been made in error, overstated, or wrongly interpreted. Be prepared to modify your hypothesis in the light of existing knowledge, but try to avoid invoking auxiliary hypotheses or exceptions to account for inconvenient observations.

- Make sure that your hypotheses are testable. This means that it should be possible to produce results that would, in principle, falsify them.
- Don't become over-attached to your hypotheses. Look for alternative explanations. Identify those that are not worth pursuing and then eliminate them. Focus on those hypotheses that provide the simplest solutions to the problem.

References

Achinstein P (1977) History and the philosopher of science: a reply to Cohen. In: Suppe F (ed) The structure of scientific theories. Illinois Books, Illinois, p 357

Allen JF (2001) Bio-informatics and discovery: induction beckons again. BioEssays 23:104–107

Lakatos I (1970) Falsification and the methodology of scientific research programmes. In: Lakatos I, Musgrave A (eds) Criticism and the growth of knowledge. Cambridge University Press, Cambridge, pp 91–196

Laudan L (1981) Science and hypothesis. D. Reidel, Dordrecht, p 183

Rapoport A (1972) Explanatory power and explanatory appeal of theories. Synthese 24:321–342

Other Useful Reading

Johnson S (2010) Where good ideas come from. Riverhead Books, New York

Chapter 6
Making Observations

Abstract The making of observations is not the primary act in research. Observations need a chosen problem context in order to be relevant. In this view hypotheses always precede observations. The most effective research is likely occur where the hypothetical bases of the data gathering and analyses are made explicit *before* data are gathered or selected.

> I am a firm believer that, without speculation, there is no good and original observation.
>
> Charles Darwin (1809–1882).

It is clear that one of the most important characteristics of science is the accuracy and precision with which observations are made. When the popular press reports on scientific progress it is often in terms that reflect the efforts and expense expended on making observations: the Hubble orbiting telescope, the equipment used to explore the ocean depths, and the Human Genome Project. The Large Hadron Collider, built near Geneva, Switzerland to test the predictions of high-energy physics, is 17 miles in circumference, and had a project budget of over five billion dollars. Much of the progress in scientific knowledge appears to be a result of the invention of better, more powerful observational devices. It would then seem to be intuitively obvious that, having identified and characterized a problem, the next step would be to start making careful observations.

It is still commonly believed that scientists use a process of generalizing from observations (induction) in order to generate the theories and hypotheses that they test in their experiments. The following is a typical explanation of scientific method based on this view (Tarbuck and Lutgens 2006):

The development of new scientific knowledge involves some basic logical processes that are universally accepted. To determine what is occurring in the natural world, scientists collect scientific "facts" through observation and measurement. Because some error is inevitable, the accuracy of a particular measurement or observation is always open to

P. Truran, *Practical Applications of the Philosophy of Science*,
SpringerBriefs in Philosophy, DOI: 10.1007/978-3-319-00452-5_6,
© The Author(s) 2013

question. Nevertheless, these data are essential to science and serve as the springboard for the development of scientific theories.

This view of the way science makes progress certainly has a respectable provenance. However, Popper (1974) explained the difficulties that attend this view. He pointed out that if we simply ask someone to "Observe", the response is almost certainly to be "Observe what?" followed by other questions, "How?", "When?", and "Why?". Of course they could respond with a catalogue of random observations, of immediate visual sensations, but these would have little meaning or purpose. Observation is always selective. It needs a chosen object, a definite task, an interest, a point of view, a problem. In this view *hypotheses always precede observations*. In the case of the human genome project the processes for generating data themselves embody complex theoretical principles, and the computerized interrogation of the database implies the application of hypothetically derived algorithms or testing procedures. The starting point for this mammoth program was a set of theoretically framed expectations and exploratory techniques. It cannot be construed simply as an exercise in data collection or the making of observations.

> In the field of observation, chance favors only the prepared mind.
>   Louis Pasteur (1822–1895).

The astronomer (probably Galileo) who first observed a double star was making observations in the light of the hypothesis or expectation that all stars were single objects. The double star represented a deviation from that which was expected. The reason why the double star was even *noticed* by the astronomer is precisely because of his hypothetically based expectations—his understanding of the structure of the heavens. His mind was prepared to notice the double star. Similarly, Alexander Fleming's seminal observation of the effect of fungal growth (Penicillium) on bacterial cultures was made in the light of his extensive knowledge of microbiology. Fleming had experience and knowledge of the distribution and shape of the colonies of Staphylococcus bacteria on the Petri dishes that he had left while he went on vacation, and it was his expectations that allowed him to recognize the importance of the observations that led to the discovery of Penicillin. It was clear to Fleming that something emanating from the Penicillium mold that had contaminated his Petri dishes was responsible for destroying his Staphylococcus colonies. The unusual behavior of the Penicillium was significant for Fleming because the identification of anti-bacterial agents was the key problem he had been working on for many years. Indeed, 6 years previously, Fleming had discovered the anti-bacterial properties of the enzyme lysozyme. He was therefore primed to make the critical observation that led to the momentous discovery of Penicillin.

The first observations of double stars and the antibiotic action of Penicillium are often referred to as discoveries, observations that were made by chance or unexpectedly. In the case of these observations there may have been no formally expressed hypothesis that was being intentionally tested by the researcher, but there would certainly have been a problem context that would have given meaning and significance to the observation. The significance of the observation, the features that made it stand out as being of interest, would be its unusualness or its being an "outlier". The observation is recognized as worthy of note because the researcher has sufficient interest and background (hypothetical) knowledge to be able to see it for what it is—an anomaly, a deviation from expectations. Such expectations are unlikely to have been made *formally* explicit, but will be an intrinsic element of the researcher's theoretical knowledge base, his experience and his interests.

We can make a distinction between observations that represent unexpected deviations or anomalies (discoveries), and those that are the product of the systematic gathering of data. The collection of data, the making of observations, may appear to be the initial response to the formulation of many research problems. For example, the production engineer who has a problem with a product dimension that is outside of specifications may start his investigations by gathering historical and current data on the variability of the critical dimension in previous production runs. He will probably also gather data on other parameters that his previous experience tells him may have affected the product dimension that is now outside specifications. These decisions to collect certain data and not others will be based on a substantial background knowledge that will include facts, hypotheses and theories. The engineer will understand that measurements in a well-controlled production line will show a normal distribution, and that these measurements maybe expected to drift over time as the equipment wears. He will have standard, theory-based methodologies for investigating the types of problems that he expects to encounter in his work. He will be looking to see if there have been any sudden changes that cannot be explained by gradual deterioration in the equipment and that may be due to some, as yet unexplained, changes in the production line. As he sets up his investigation and starts to gather data, he will be guided by hypotheses that *may* or may not be explicitly formulated.

Similarly, a lepidopterist studying the problem of environmental changes on a moth population may well start by a survey of the moth species present in the selected area. He will set up lamps at night to attract moths that will be captured, identified and counted. In this case the lepidopterist's hypothetical knowledge base will include the recognition of the effect of local weather conditions on the distribution of moths and the seasonal effects on the species present and their numbers. Again, the hypotheses being tested may not be explicitly formulated. However, in both of these examples it is clear that a clear statement of the research problem, and a clarification of the hypotheses that form the context for the observations being undertaken will enhance the likelihood of the research being successful. Such research is more likely to succeed for the following reasons:

1. A clear statement of the hypothesis to be tested will make it more likely, once the study has been completed, that observations will have been made in the right place, and in the right numbers. There is nothing more frustrating for the researcher than to having to repeat a study because the number of observations made was insufficient to allow the appropriate statistical tests to be done.

2. The initial framing and screening of hypotheses will allow the elimination of hypotheses that do not stand up to initial scrutiny, and that are not worthy of experimental test.

A good example of the importance of the framing of hypotheses prior to the collection of observational data is provided by Darwin's collection of finches from the Galapagos Islands on the second voyage of the Beagle (Darwin 1857). When he was collecting the specimens of the famous finches that provided such a definitive example of the process of Natural Selection, Darwin had not yet fully formulated his theory. Upon returning to England, his ideas began to crystallize, and he realized that different islands were home to different species of birds. However, he had not labeled the finch specimens by island. His data (observations) were therefore useless as support for his ideas. Fortunately, others on the Beagle, including Captain FitzRoy, had also collected specimens, and they *had* allocated the species to islands. Darwin was accordingly able to use the distribution of different species of finches as evidence to support his theory.

The technique of *participant observation,* which is widely established and used in the field of anthropology, might be seen as a counter example to the view that the framing of hypotheses should precede the making of observations. According to this technique the researcher, unburdened by theoretical preconceptions, receptive without interfering, immerses himself or herself in a new culture, participating in people's daily lives and recording what is seen and heard. Through observation, the anthropologist collects data for subsequent analysis. The method of immersion allows the researcher to establish close relationships with the subjects of the study. It is claimed that the more the researcher becomes familiar with people's lives, the more he or she can effectively read the meanings of the data collected. This ability leads the anthropologist to draw reliable conclusions from the research.

However, consciously or unconsciously, the observations made by the researcher will be affected by his interests, his knowledge of this or other cultures, and by his preconceptions or prejudices. There may be phenomena that he may not observe at all unless he has some hypothetical framework or expectations that will allow him to see them. For example, the celebrated anthropologist Margaret Mead may have been misled in her studies of Samoan adolescent sexuality because she did not recognize the propensity of the girls in her study to engage in playful deceit (Freeman 1999). It is evident that the anthropological researcher must acquaint himself with the current state of knowledge in his field, and be aware of the beliefs and preconceptions that he brings to his work in order to make salient and relevant observations. Indeed, truly interesting observations are more likely to be made as the anthropologist develops a hypothetical framework of understanding of his subject matter, for it is then that anomalous observations or deviations may be noted, observations that potentially challenge current theory within the field of knowledge.

Now, it is widely accepted that there may be times when a researcher may actually proceed by examining data in order to discover patterns or anomalies that may lead to new theories or hypotheses. The accumulation of huge quantities of data because of commercial, social and scientific (especially medical) record keeping, and the rise in computing power has resulted in the development of techniques for the interrogation and analysis of databases referred to as "data mining". The investigator may look for clusters, associations or correlations within the data. Such techniques have proved to be particularly useful within the retail industry where analyses of customer purchasing preferences have been used to develop products and promotions that appeal to specific customer segments. However, it should be noted that even the setting up of a database requires decisions with respect to the data to be collected, and this will be based upon some hypothetical knowledge of the problem. The mining techniques themselves will be based on theoretical preconceptions about the most likely strategies for identifying useful patterns or relationships. It is worth reiterating that the most effective use of such research techniques is likely occur where the hypothetical bases of the data gathering and analyses are made explicit *before* the data are gathered or selected.

Consider once more the case of the data generated as a result of the Human Genome Project. Mining of these data is done by identifying sequences of DNA that code for proteins and then comparing them with sequences for genes whose functions have already been established. Other approaches include looking for patterns and relationships between the data. However, as the sheer scale and complexity of the human genome has become manifest, it is clear that it will be a huge intellectual challenge to generate the hypothetically-based algorithms required to make sense of the data. Similarly, the Deep Field images generated by the Hubble telescope provide a wealth of data to be interpreted in the light of new theory. As telescopes probe deeper into the history of the universe, it becomes necessary for observations to be done in the infrared range to accommodate the extreme red shifts which theory predicts will be seen at the edges of the universe. Once again we are becoming aware that the data themselves do not give up their secrets, and we see that the creative generation of hypotheses and subsequent testing of them against the data is ultimately the route to new knowledge.

Practical Points

- Recognize that the making of observations is not the primary act in research. Whether doing a literature search or "gathering data" make sure that you clarify the underlying expectations, assumptions or hypotheses that inform your search.
- Evaluate potentially competing hypotheses before making a decision about which ones are worthy of being tested by the gathering of observations.
- Make a clear and explicit statement of the hypothesis to be tested *before* setting out to make observations. Ensure that the observations will provide sufficient data to allow a rigorous test of your hypothesis.
- Check your choice of observations to be made. Make sure that there are no other key observational parameters that should be measured and recorded.

- Prepare yourself to take appropriate notice of unusual or anomalous observations. Don't simply ignore, dismiss or explain away observations or data that do not fit within your current understanding.
- Whether your observations confirm or refute your hypothesis, check that your underlying thinking is secure. Ensure that recording biases, errors in equipment, or any other observational errors have not compromised your observations.

References

Darwin C (1857) Letter to A. R. Wallace, 22 December 1857, vol 6. In: Burkhardt F, Smith S (eds) (1987) The Correspondence of 1844–1846, p 514

Freeman D (1999) The fateful hoaxing of Margaret mead: a historical analysis of her Samoan research. Westview Press, Boulder

Popper KR (1974) Conjectures and refutations. Routledge and Kegan Paul, London, p 46

Tarbuck E, Lutgens F (2006) Earth science. Pearson Prentice Hall, New Jersey

Chapter 7
Seeing is Not Necessarily Believing

Abstract The instruments we use to test our hypotheses *themselves* embody hypotheses, and consequently they embody the uncertainties that are an integral element of all hypotheses. We need also be aware of the theoretical constraints that are embodied within the devices we use to make our measurements and observations of the world.

> Of all things the measure is Man, of the things that are, that they are, and of the things that are not, that they are not.
>
> Protagoras (ca. 490–420 BCE).

In "The Structure of Scientific Revolutions", Thomas Kuhn describes a fascinating experiment reported in 1949 by the perceptual psychologists Bruner and Postman. They briefly flashed images of playing cards onto a screen and asked their subjects (Harvard students) to identify them. They found that when subjects were shown anomalous playing cards, for example, a black four of hearts, they reported seeing their normal counterparts, in this case a red four of hearts. After repeated exposures, most subjects realized that something was wrong and eventually they described the anomalous cards correctly. However, in some cases, the subjects were unable to identify such cards even after repeated exposures, and they became confused or distressed (Kuhn 1996). It was clear that the students' ability to recognize the anomalous cards was determined by their preconceptions about the nature of playing cards. Expectations based on our beliefs or understanding of the world will affect our ability to observe it. The act of observation is not simply a matter of directing our gaze or listening carefully. Rather, it seems clear that expectations based on our beliefs or understanding of the world will affect our ability to observe it; our observations are theory laden. This will apply to observations made with our senses, but also to those made with observational devices.

Consider the following example. Most of us are familiar with the simple spring scale, in which the weight of an object may be measured by the extension of the

P. Truran, *Practical Applications of the Philosophy of Science*,
SpringerBriefs in Philosophy, DOI: 10.1007/978-3-319-00452-5_7,
© The Author(s) 2013

spring on which it is loaded—the longer the spring, the greater the weight. The valid operation of the instrument is based on a theory called Hooke's Law that describes the relationship between spring extension and the force (weight) applied. The Law also stipulates that this relationship is proportional as long as the load does not exceed the elastic limit of the spring. The spring scale will not give valid weight readings if the load is too heavy because the spring will be stretched beyond the point where it will return to its original length. On the other hand, for small loads the extension of the spring will be insufficient to give sufficiently accurate or precise results. Accordingly, there is a restricted range within which the instrument may be used appropriately, and anyone using a spring balance will need to understand these limitations. An observation of weight made with a spring scale will therefore embody the theoretical considerations that are described by Hooke's Law. This is true of any observations—they are all laden with theoretical content. A thermometer will embody the theory about the regular expansion of mercury with increasing temperatures over a defined range, and a radio telescope will embody theories about the focusing of radio waves onto an antenna, their electronic amplification and ultimately their visualization on a computer screen. However, this is not just true of observations made with instruments and measuring equipment, it is also true of the most direct observations that we can make— with our eyes, our ears or any of our senses.

Intuitively we believe that if we have seen something with our own eyes then it must be true. Yet most of us are skeptical about so-called eyewitness reports of flying saucers or ghosts, even if such phenomena have been seen at first hand by people of good character who swear that they are telling the truth. We are likely to attribute such reports to observational errors, to tricks of the light, to wishful thinking or to willful deception. The notion that there are simple, pure observations or data that form the bedrock of our understanding does not stand up to scrutiny. We are acutely aware of the inadequacy of our own eyes as observational devices. What we see, or don't see, will be directly affected by our expectations and our experience. This is the territory of the professional conjuror and the crime scene investigator. The conjuror takes advantage of our preconceptions about how playing cards may be manipulated in order to fool us into believing that our chosen card has moved from the middle to the top of the pack; the detective will be aware of how suggestible witnesses may be, and how their observations of a crime may be influenced by their prejudices.

Even if we make the simplest, most straightforward of visual observations, what we see will be determined by the capability of our eyes and the neuronal circuitry in the brain that interprets the raw signals generated by light falling on the photoreceptor cells of the retina. Humans cannot see light that has a shorter wavelength than that of blue light and longer than red light. We are not very good at seeing at night. These limited capabilities were presumably adequate for our prehistoric forebears who hunted and gathered by day and sought shelter at night. They were probably inherited from our hominid and pre-hominid evolutionary ancestors. Our eyes evolved to meet the challenges of survival in the plains of Africa and they represent one element in the package of capabilities that provided a human solution to the problem of finding food, killing prey and avoiding predators. We can therefore regard the human eye and its visual cortex as embodying a hypothesis, a solution to the problem of detecting electromagnetic radiation in what we call the visible range. The hypothetical solution to the problem of seeing developed by humans was not developed or designed by anyone. It is a solution that evolved by the process of natural selection in response to the problems of survival encountered by humans and, of course, the long line of ancestors that preceded them.

A range of different theoretical solutions to the problem of seeing has been developed during the course of evolution. Bees, with compound eyes made up of hundreds or thousands of units (ommatidia), each with its own individual lens and detector, use a theoretical solution developed by insects. The eyes of the bee have quite different visual capabilities to those of the human. Bees cannot see as clearly as humans, but they are very good at detecting movement, and they can see into the ultra-violet range of electromagnetic radiation. The bee is able to see flower tracks, visible in ultra-violet light, on the petals of certain flowers that will lead it directly to the source of nectar. The visual capabilities of the bee have co-evolved with the patterns of coloration of the flowers on which it feeds and which it pollinates. Its eye embodies a hypothetical solution to the problem of detecting the sources of nectar and pollen on which it feeds.

The hypothetical elements incorporated by evolution into the structures of the human eye and the visual cortex are supported by a framework of knowledge about the world that each of us will have developed as a result of our experience. This visual knowledge will have started developing early in infancy and continues throughout life. We may have developed a theoretical visual knowledge of playing cards that is very useful for playing bridge, but it makes it difficult for us to immediately recognize anomalous cards. Similarly, when most of us look down though a microscope for the first time, we cannot make sense out of anything that we are seeing. If we are asked to observe what we can see in a drop of pond water we may see only meaningless blobs and spots. It takes experience and instruction for us to be able to recognize that the objects swirling around in our field of vision are cells, or tiny plants and animals. We need to build up a hypothetical knowledge of the microscopic world before we can interpret the images that fall on our retinas. The idea that there may be microscopic organisms visible in pond water (a notion that did not exist before the invention of the microscope) is necessary before we are able to see them.

Just as a barometer is an instrument used for observing air pressure, and a trawl net is an instrument for gathering data on marine life, so we can regard the eye as an instrument for observing or gathering data on electromagnetic radiation. Note that a questionnaire or an interview can also be regarded as an instrument for making observations or gathering data. It is clear however, that no matter the nature of the instrumentation being considered, it is vital for the researcher to be aware of the hypothetical assumptions that underlie the use of the equipment or instrumentation being used to make observations. Anyone using data derived from visual observations will need to be acutely aware that all observations will be colored by expectations, or by a theoretical perspective. Similar care will need to be applied in the application of certain psychometric tests that will have been developed and calibrated in a particular culture, and may not be applicable in different circumstances. The correct interpretation of psychometric test results will depend upon a good understanding of the structure of the test itself (Mann 1985).

The professional photographer is acutely aware of the theoretical constraints of the camera and detectors that he is using. The size of the photo-detector array in the digital camera will affect what is seen in the final image, as will the interplay between aperture and exposure. In the case of sophisticated scientific instruments there may be complex theoretical considerations involved in the making of an observation. The exquisitely detailed electron micrographs of the fine structure of the cell are made by processes that involve staining specimens with heavy metals, embedding them in plastic, and then making ultra-thin slices (sections) before directing a beam of electrons through them on to a fluorescent screen. Because electron micrographs are snapshots of ultra-thin slices through the cell, the images will only be meaningful in the light of the theoretical constructs made by the cell biologist. The biologist makes links between the fine structure of the cell seen in the electron microscope and those seen in a light microscope. His knowledge of cell structure, derived from the less detailed images seen in the light microscope, will allow him to construct hypothetical pictures or models, that will enable him to understand the images seen in the electron microscope. Their interpretation will involve judgments about whether the structures observed in the image are artifacts of the preparation and imaging processes, or whether they truly reflect the microstructure of the cell. It may be difficult to determine whether totally new structures seen in an electron micrograph are genuine objects or artifacts. Increasing confidence in such judgments will be built up as observations are reliably repeated, as similar images are seen using different methodologies and as new claims are subjected to the processes of critical discussion and peer review.

In more complex cases, such as the use of functional magnetic resonance imaging (fMRI) to detect brain activity in real time, it may be even more difficult to interpret instrumental observations. The scientist observing the screen during an fMRI session will see a pattern of colored areas corresponding to different parts of the brain responding to different stimuli. These patterns will have been produced by the action of magnetic pulses that cause protons in biological materials to produce radio waves. Certain materials in the body will produce radio waves that decay less rapidly than others and this allows the brain activity to be visualized using the appropriate computer software and imaging equipment. Given the huge amount of theory (electromagnetic, biological and electronic) that underlies these observations, how do we know that they correspond with any reality? The key thing here is that the images made using equipment with complex theoretical underpinnings will have been developed, and evolved from, more simple equipment—much in the way the complex electron micrograph has developed from the images seen with the light microscope—and this gives us justifiable confidence that what we are seeing on the screen corresponds with real phenomena in the brain. The danger arises when operators do not really understand the underlying assumptions and inherent limitations of such complex, sophisticated technology.

Any instrument used for making measurements in research will have limited capabilities and these will be based upon theoretical limitations embodied in its design. In the case of physical instruments there will also be variations between individual units, and the performance will change with age due to wear or deterioration of the mechanical and electronic components. Scientists and non-scientists will often make use of commercially developed analytical equipment that, although it is developed to high standards, may be used inappropriately. This may happen when the operators do not understand the theoretical basis of the analytical system being used. Concerns have even been expressed about the use of digital photography and image manipulation in cell biology. Unless the scientists or technicians using them have a thorough knowledge of the programs they are using for image manipulation, they are "performing "black-box" image enhancements that they do not control to any significant degree" (Greene 2005).

So what attitude should the researcher hold towards the observations and data that he makes use of in the testing of hypotheses? Consider the words of Protagoras quoted at the beginning of this chapter, "Of all things the measure is Man, of the things that are, that they are, and of the things that are not, that they are not." I take this to mean that Man constructs his world—a world of guesses, hypotheses and theories. Moreover, Man also constructs his observations. Observations may be made with instruments that we have inherited as flesh and blood creatures, our eyes, ears and noses. We need to be aware of the in-built, theoretical limitations of these natural instruments, limitations that are both inherent and the results of our experience. We need also to be aware of the theoretical constraints that are embodied within the devices we use to make our measurements and observations of the world. The instruments we use to test our hypotheses themselves embody hypotheses, and consequently they embody the uncertainties that are an integral element of all hypotheses. The words of the Pre-Socratic philosopher Xenophanes ring true: "For all is but a woven web of guesses" (Popper 1974).

Practical Points

- Maintain a healthy skepticism towards observations, whether they are made by ourselves us or by someone else. Be wary of the effects that preconceptions or expectations can have on what we see or measure.
- Understand fully the theoretical basis and assumptions behind the equipment and instruments (whether physical or conceptual) that you are using.
- Be sure that the theoretical assumptions that are made when we are using our instruments are fully justified and validated. A biochemical test developed for use with one medium (for example blood) may inappropriate for use in another

(urine); a psychometric test developed for one culture may give biased results when used in another.

- Familiarize yourself with the principles of equipment and test validation. These validations will provide a sure foundation on which to make observations in which you can have confidence.
- Ensure that any test instrument is used within the limits for which it was intended. This will include the range over which it performs with acceptable accuracy and precision. Do not accept and use test systems reported in the literature without validating their performance yourself. Validated, commercially produced test systems and instruments may perform differently in your field or laboratory environment.
- Use experimental designs that take into account the limitations of the tests systems being used. For example, if the test instruments produce poor precision, then greater numbers of repetitions will need to be incorporated into the experimental design.

References

Greene MT (2005) Seeing clearly is not necessarily believing. Nature 435:143
Kuhn T (1996) The structure of scientific revolutions. University of Chicago Press, Chicago, 63
Mann AH (1985) The clinical relevance of psychometric testing. Eur J Clin Pharmacol 28:31–34
Popper KR (1974) Conjectures and refutations. Routledge and Kegan Paul. 153

Chapter 8
Testing Hypotheses

Abstract There are useful guidelines that allow us to identify effective and powerful ways of testing hypotheses. Many of these are informed by an attitude towards hypotheses and theories that views them with caution and skepticism. The community of researchers is constantly striving to make itself aware of which critical strategies are being employed to achieve a rigorous approach to the framing and testing of hypotheses.

> The growth of knowledge, by and large, has to do not with replacing or amending some well-confirmed theory, but with testing specific hypotheses in such a way that there is a good chance of learning something…
>
> Deborah Mayo (1996).

Testing our hypotheses allows us to distinguish between competing solutions to our problems. We have already seen that the way in which we frame a hypothesis will determine how testable it is. It should be plausible, internally consistent (clearly framed) and potentially falsifiable i.e. when we set out to gather data to test our hypothesis there should be a possibility that the data we collect may *disprove* it.

Once we have formulated our hypotheses, how should they be tested? The view that science follows clear rules for the appraisal of theories was challenged by Thomas Kuhn, and subsequently by Feyerabend (1975). They asserted that there are no methodological rules that are always used by researchers. However, there are some useful guidelines that allow us to identify effective and powerful ways of testing hypotheses. These guidelines assume a scientific attitude towards hypotheses and theories, one that views them with caution and skepticism. The community of researchers is constantly striving to make itself aware of the strategies employed to achieve rigorous framing and testing of hypotheses. Once the preliminary process of critical assessment of competing hypotheses has been completed, the researcher will arrive at a point where an experiment may be considered. A good experiment allows the researcher to establish if the hypothesis can be

P. Truran, *Practical Applications of the Philosophy of Science*,
SpringerBriefs in Philosophy, DOI: 10.1007/978-3-319-00452-5_8,
© The Author(s) 2013

corroborated (survive a severe test), or whether it is to be rejected and alternative solutions sought. On the other hand, a poorly conducted experiment may lead us into error, that is, the rejection of a hypothesis that is in reality correct (a Type I error), or conversely the acceptance of a hypothesis that is false (a Type II error). The sources of experimental error are manifold, but well crafted, robust experimental designs provide the means by which we minimize the chances that we are in error.

Experiments allow us to establish causal relationships, for example, that smoking of tobacco is a cause of lung cancer, or that the gravitational pull of the earth will cause the deflection an asteroid from its orbit around the sun. The causes and effects investigated in an experiment are usually called variables, factors that can take different values. For example, in the work done to investigate the cause of lung cancer, smoking was the postulated cause (or independent) variable, and could vary simply as smoker or non-smoker, or as tobacco consumption per day. The effect (or dependent) variable was the lung cancer that may or may not have developed in a particular subject. Because a particular effect (in this case, lung cancer) can have so many possible causes (genetic effects, diet, lifestyle as well as smoking) it can be difficult to establish a firm relationship between an effect and its cause or causes. For this reason, we usually carry out experiments by investigating a limited number of variables in order to simplify things. Investigations are conducted in such a way as to eliminate or reduce the effects due to other variables that might confuse or confound the relationship between the variables of interest.

The famous epidemiological study conducted in 1954 by Doll and Bradford Hill reported on the rates of cancer and other diseases in doctors of medicine, and related this to their smoking habits (Doll and Hill 1954). Although there was some evidence of an association between smoking and lung cancer, it appeared that other factors might play an important causal role. By confining the study to one class of subjects (doctors), differences due to other causes, such as working environment, diet or standards of living, were effectively eliminated as alternative potential explanations for the pattern of the disease. In this way the design of the study *controlled* for the effects of other variables in order to provide convincing evidence of the link between the two variables, smoking and incidence of lung cancer. These studies on smoking and cancer were later backed up by laboratory experiments in which extracts of tobacco smoke were applied to animals under controlled conditions to establish the relationship between the tobacco smoke and the development of cancer. Such studies could not, of course, be carried out on human subjects for ethical reasons, and the mammals used for the laboratory experiment were believed to provide a good model for the development of the disease. The work done on smoking and lung cancer had all the key elements required of good experimental science: variables that could be measured or categorized, the control of potentially interfering variables, and replicability—the ability of work to be reproduced by other researchers.

It is important to realize that experiments are not primarily instruments of discovery. It is true that an experiment may lead to an unexpected result, or totally novel observations. A good example of this type of experiment is the discovery of X-rays in 1895. Röntgen was working in his laboratory with a cathode ray tube that produced a fluorescent glow and was shielded with black paper. During the course of this experimentation he noticed the glow of a cardboard screen placed outside his apparatus. The screen had been painted with a fluorescent material, and Röntgen realized that rays from the cathode ray tube had passed through the black paper and caused the material to glow. It is worth noting that Röntgen's discovery was not due to simple luck. In fact he had already planned and was about to execute an experiment that would have revealed the existence of X-rays. Nevertheless it is true that Röntgen did not anticipate or expect the results of this experiment with the cathode ray tube. Similarly, the results of high-energy experiments carried out in a cyclotron or any other particle accelerator may not be known or anticipated in advance. Like expeditions to the sea floor, or to polar regions, they are really attempts to generate or identify phenomena that will provide a source of new problems, and subsequently be subjected to the processes of hypotheses generation and experimental test.

> ... scientific work of an experimental or exploratory character starts with some expectation about the outcome of the enquiry. This expectation one starts with, this hypothesis one formulates, provides the initiative and incentive for the enquiry and governs its actual form. It is in the light of this expectation that some observations are held relevant and others not; that some methods are chosen, others discarded; that some experiments are done rather than others.
>
> Peter Medawar (1969)

The results of a good experiment should not normally be a surprise. The researcher should have a clear idea of the results expected *in advance* of the implementation of the experiment. If the data are to be evaluated in graphical terms, it should be possible to produce a rough graph of the results *before* the experiment is conducted. This seems to fly in the face of our intuitive understanding view of how an experiment should be conducted, and it may be surprising to many aspiring researchers who consider that it is a form of cheating to have a clear idea of our experimental outcome. Surely, if the researcher anticipates the results, will there not be an overwhelming temptation, either conscious or unconscious, to influence them, to massage or skew the data, even if only unintentionally? On the contrary, unless the researcher makes the expected results as explicit as possible there are clear dangers of falling into error, for it is much easier to avoid these dangers if we understand where they lie.

Consider, as an example, a simple clinical trial comparing the effects of a treatment on the levels of a hormone in two groups of subjects, a test group and a control group. There will need to be a careful consideration of the potential variability associated with the subjects of the trial, and the sources of random error (noise) or systematic error (bias) due to the measurement instruments to be used and their operators. Such an analysis is essential for the identification of an appropriate experimental design. For a start, if we are going to compare the mean hormone level in the two groups we will need to have, *in advance,* a reasonable idea of the difference in hormone levels expected and of the likely variability of the measurements to be made. If the difference between mean levels is expected to be small, and the measurement variability is expected to be large, then the numbers in each group will need to be large in order to allow a statistically significant difference to be demonstrated. Conversely, a larger expected difference with greater measurement precision will indicate smaller numbers in each test group. This assessment will allow us to determine the number of samples to be taken, and the number of replicate measurements to be made. There may be a need to randomize samples to take into account systematic variation or drift in the measurement systems used. The identification of other sources of error, such as researcher or operator bias, will determine how subjects are chosen, how samples are collected and identified, and whether a double blind trial is called for. Consideration of the results expected therefore allows the researcher to minimize the chances of falling into error. Incidentally, this is particularly important in the case of clinical trials, where poor experimental design may lead to equivocal results. There may be inconclusive, but potentially important, evidence derived from the trial outcome, for example indications of unacceptable side effects. From an ethical perspective, the mere possibility that such side effects may be real will make it difficult to repeat the experiment with a better design. It will be vital in such circumstances to get the experimental design right. There may be no second chance.

At this point, *before* implementing the experiment, the researcher should be speaking to the statisticians and experts in experimental design. Good advice on design will not only ensure that the scale of experiments will be appropriate for achieving clear, unambiguous results, but will also establish if there are opportunities for looking at the relationship between more than two variables in a single experiment. The use of experimental designs based on factorial designs or orthogonal arrays can allow the researcher to investigate the relationship between multiple variables in a single experiment. In contrast to experiments that consider only one pair of variables at a time in a step-by-step fashion, factorial designs can provide crucial information about interactions between different variables. Unfortunately, the use of full factorial designs will lead to the exponential increases in the sizes of experiments as more variables are considered. However, the careful use of appropriate reduced or fractional factorial designs will allow the researcher to make judgments about the variables and interactions to be investigated. Such strategies can dramatically improve the throughput and efficiency of experiments, and can be particularly useful in screening experiments done to identify the key variables that will be the subject of subsequent, more intensive, larger scale experiments. Reduced factorial designs are widely used in industry for investigation and optimization of complex processes that have a large numbers of variables and potential interactions between those variables, and they are a well-established experimental tool in the agricultural and social sciences. They are less widely used within traditional research fields, perhaps because the statistical analyses required are more complicated, but their use should be considered wherever large numbers of variables are being investigated, particularly in research method development.

Well-designed and carefully executed experiments will provide the best opportunity for appraising our hypotheses. The results may allow us to conclude that we are on the right path to the truth. On the other hand we may be obliged to accept that our hypothesis is wrong, an error, or inferior to alternative hypotheses. This type of error, the mistaken hypothesis identified by a well thought-out, robust experiment, differs from the errors of indeterminacy and bias that bedevil the experimenter. The clear identification of a mistaken or rejected hypothesis is an error from which we can learn. We will know one area where the answer to our problem probably does not lie, and we will be confident that we are correct in rejecting our hypothesis and moving on.

> We have a habit in writing articles published in scientific journals to make the work as finished as possible, to cover up all the tracks, to not worry about the blind alleys or describe how you had the wrong idea first, and so on. So there isn't any place to publish, in a dignified manner, what you actually did in order to get to do the work.
>
> Richard Feynman (1918–1988) (Feynman 1965)

Finally I would like to say a little about the attitudes of students towards theory and experiment. The portrayal of science in high school is skewed very much towards the acquisition, demonstration and utilization of theoretical knowledge. Experiments rarely represent an opportunity for testing of hypotheses, but typically take the form of demonstrations of theoretical principles or stage-managed devices for "discovering" them. There is very little opportunity for students to learn about the great experiments and the great experimenters. There will be limited insight into the inventiveness, persistence and technical creativity that are necessary to carry out an important experiment, and the ways in which a well-executed experiment can be the engine for development of theoretical ideas. The accounts of classic experiments are often written in archaic language and type, and the diagrams of experimental equipment may be difficult to understand and appear to be outdated and inaccessible.

Unfortunately, when students start to explore the reports of experiments in more recent scientific literature they will discover that they provide very little indication of the real processes of experimentation. As Richard Feynman has said, there is no place where scientists can provide an account of the iterative dialogue between ideas and critical test, of the mistakes, errors, and the fine-tuning that take place along the road to achieving an acceptable experimental result. Scientific reports and papers published in learned journals appear as sanitized accounts that, as pointed out by Peter Medawar (1963), distort the reality of the scientific process. For this reason, when apprentice researchers begin to learn about the experimental techniques used within their field, they will depend to a large extent upon the experimental techniques taught at the undergraduate level and those learned from laboratory colleagues. The apprentice researcher will become immersed in an experimental tradition established within a department and a field of endeavor. This is fine as far as it goes, but there is much to be recommended in reading accounts of actual research practice and attitudes, especially when these have been written by practicing scientists. This will broaden the researcher's understanding and appreciation of the thinking of successful scientists, the creativity and rigor they bring to bear on the resolution of problems, and the sheer persistence required to address the technical difficulties that may occur in pursuing experimental programs. This understanding will help the student researcher to appreciate the value of a well-conducted experiment that yields a negative result. Such experiments will tell us where the answer does *not* lie, and provide a basis for the fine-tuning that will eventually take us in the right direction. The history of science can be seen very much the history of error, for it is by learning from our mistakes that we make progress.

Practical Points

- Learn from the great researchers. Read about the great experiments, even if these have not been in your field. Try to find out about the realities of research practice and attitudes by reading and talking to other researchers.
- Make sure you have a clear idea of the results that you expect from your experiments. If you expect to be able to represent the results of an experiment as a graph, sketch out the expected graph *before* you start to design your experiment. If the results will be numerical, do a preliminary estimate of what you expect the numbers to be. This is not cheating!
- Learn about the design of experiments. A well-designed experiment will reduce the likelihood that you will get equivocal results or will have to repeat your work. The use of multifactorial experiments can save you huge amounts of time, money and frustration. Consider the application of reduced factorial designs or orthogonal arrays for screening or optimization experiments.
- Consult statisticians during the design of your experiment—not just as an afterthought. Apart from anything else it will improve your relationship with these crucial experts. Remember that statisticians will be able to advise you on the design as well as the analysis of your experiments.
- Be on the lookout for unexpected results. The better the design and execution of your experiments, the more significant such surprises are likely to be.

References

Doll R, Hill AB (1954) The mortality of doctors in relation to their smoking habits: a preliminary report. British Medical J, 1(4877):1451–1455

Feyerabend P (1975) Against method. NLB, London

Feynman R (1965) The development of the space-time view of quantum electrodynamics. Nobel Lecture

Mayo D (1996) Error and the growth of experimental knowledge. University of Chicago Press, Chicago and London

Medawar P (1963) Is the scientific paper a fraud? The Listener Sep 3: 377–378

Medawar P (1969) Induction and intuition in scientific thought. American philosophical society (These essays are transcripts of the Jayne lectures given by Peter Medawar in 1968. Winner of the Nobel Prize for Medicine in 1960, Medawar brings the perspective of the seasoned researcher to these expert and accessible reflection on the nature of scientific thought. Clear, concise and an excellent read.)

Chapter 9
More on Induction and Justification

Abstract There is no logical process by which science moves from observations to testable hypotheses. Rather, the progress of science depends upon the creative making of informed guesses and then subjecting them to criticism. We must cultivate the highest critical standards towards our own work so that we can identify and eliminate error as efficiently and rigorously as possible.

> When propounding a general theory in science, the one thing one can be sure of is that, in the strict sense, such theories are mistaken. They are only partial and provisional truths which are necessary... to carry the investigation forward; they represent only the current state of our understanding and are bound to be modified by the growth of science...
>
> Claude Bernard (1813–1878).

When the power of scientific thinking and methodology first became evident it seemed that science provided a reliable road to the truth about the natural world. Early philosophers of science proposed that the power of scientific laws lay in their being generated from a bedrock of empirical observations—observations made directly from the senses, by experience and by experiment. They believed that the scientist derives hypotheses, the patterns or generalizations that characterize scientific knowledge, from these clear unbiased observations. This process is called induction, and it was claimed that it provided both an explanation of the way that scientific theories may be *justified*, and the ways in which they are *discovered*. The following is a simple illustration of how induction was believed to yield a scientific hypothesis. I may see Monarch butterflies appear every spring in Minnesota and disappear every fall. Every time I go to Mexico I see the butterflies appear in the fall and disappear in the spring. By a process of induction, I can infer that the Monarch butterflies are performing an annual migration. My inference is justified by the evidence of my own eyes and by the logical reasoning that I have applied to the repeated observations of the butterflies.

P. Truran, *Practical Applications of the Philosophy of Science*,
SpringerBriefs in Philosophy, DOI: 10.1007/978-3-319-00452-5_9,
© The Author(s) 2013

Francis Bacon (1561–1626), who described and elucidated the idea that science makes progress by the process of induction, recognized that passive scientific observations could not provide the sole basis for growth of scientific knowledge. We could spend a lifetime making observations of the world without seeing the production of fire by the rubbing together of two sticks, the generation of a light spectrum by a prism or, indeed, the appearance of Monarch butterflies in Mexico. In order to enrich the quality of our experiences Bacon advocated the performing of "experiments", what he called "histories". In the Baconian view, the scientist experiments by trying things out, by a kind of play that will allow nature to reveal itself. For example he might explore the properties of light by observing the patterns of refraction in prisms of different shapes and compositions, from different sources or in different media. It is this richer expansion of our repeated observational experience that, in the inductive view, will lead to understanding and the truth. The inductive view of science became the standard way of explaining how science makes progress and provided a rational for the power and apparent truth of science, particularly in the wake of the enormous success of Newtonian physics. It was endorsed by Newton himself as the method of the sciences, and in the 19th century was further elaborated by John Stuart Mill.

It was the Scottish philosopher David Hume who, in the 18th century, upset the apple cart of induction by arguing logically that no number of repeated observations could ever provide the proof or justification of a proposition. Repeated observations can never provide proof of a general statement such as "All swans are white". No matter how many times we confirm this proposition by an observation of another white swan, there may always be a future observation of a non-white swan. In fact this happened when Europeans first colonized Australia and saw black swans for the first time. The same argument may be extended to the daily rising of the sun. Our past experiences of sunrise can never provide absolute proof that it will happen tomorrow. Hume argued that the validity of induction depends upon the belief in the uniformity of nature. However, this belief could not be logically justified except by an appeal to the principle of induction itself—a circular argument. In response to Hume's critique, it was pointed out that the repeated observations of events would give us a very high level of confidence that they would continue in the future. In practice we do not doubt that the sun will rise tomorrow, and arguments have been made to justify induction by appeals to probability and confidence. Nevertheless, the impact of Hume's analysis (generally called Hume's problem) has been devastating to the view that the truth of science is based on the quality and quantity of the observations upon which its generalizations about the natural world are made.

Hume's critique of induction may seem to be a trivial issue that only philosophers could possibly get exercised about. Hume himself backed off the full consequences of his argument and the extreme skepticism that it implied. He advocated a common sense approach in which we should accept induction even if it could not be logically justified. No one is really going to order his or her life around the possibility that the sun may not rise tomorrow. Nevertheless, Hume's problem has hugely affected both the ways that we view the truth of science and our understanding of the methods by which scientific knowledge grows. At a stroke, Hume demolished the idea that the truth of scientific propositions could ever be justified by induction from our observations and the results of our experiments. The comfortable notion that there was a secure route to scientific truth, or indeed, that it was even *possible* to attain certain truth about the natural world, was gone.

So what then is the means by which science makes progress, and if science does not provide certain knowledge, what does it tell us? If the argument against induction casts doubt on the accepted view that observation is the generative act in the process of scientific discovery, how may scientific knowledge be justified? The 19th century philosophers William Whewell and Charles Peirce argued that, rather than being based on the certainty of careful observations, scientific progress requires interplay between the creative processes of imaginative generation of hypotheses, and their empirical testing by observation and experiment. In Whewell's opinion no general statement, not even the simplest generalization, can arise merely from the collection and putting together of raw data. There is no logical process by which science moves from observations to testable hypotheses. Rather, the progress of science depends upon the creative making of informed guesses that might explain the phenomenon under investigation. Once we have made our hypotheses, these may be subjected to criticism. These processes of criticism are, in contrast to those of hypothesis generation, subject to the procedures of logic. We can make logical deductions (or predictions) from our hypotheses, and these can be tested empirically by making observations or conducting experiments. This scheme for the way science makes process is called the hypothetico-deductive model.

Magee (1985) gives an excellent illustration of the power of the hypothetico-deductive approach and the differences from the attitudes and research strategies based on induction. Let us suppose that we believe that it is a law of nature that water boils at 100°C. We may start off with repeated measurements to inductively verify the boiling point. Of course, none of these repeated measurements would provide *proof* of the boiling point of water as being 100°C. Even worse, all these confirming instances would never give us reason to doubt, let alone replace our original statement. However, we may decide to try a different approach—to investigate if there are circumstances in which this statement does *not* hold. If we have any imagination we will quickly discover that at higher altitudes the boiling point is lower; in closed vessels it is higher. Now, says Magee, at this point we could take a wrong turn. We could salvage the law by making it narrower, by revising it to say that water boils at 100°C, but only if this take place in open vessels—and the measurements have to be done at sea level. And we could proceed in a similar vein to pin down our knowledge of the boiling point of water. However, by doing so we would be missing a great opportunity to discover a new problem that could be framed as follows: "Why does water *not* boil at 100°C in all circumstances?"

This new problem will challenge us to produce a hypothesis that is richer and has greater explanatory power than our original simple statement. This new hypothesis may allow us to identify the relationship between pressure and boiling point so that we could calculate boiling points under different circumstances. We would then start to look for circumstances that would challenge our new hypothesis so that we could improve it even further. Thus, by a process of making bold conjectures and testing them until they break down we are able to advance our knowledge. Our knowledge cannot grow if we only search for confirming instances, unless we accidentally stumble upon a counter example (a deviation). When this does happen it will have been the best thing that could happen to us, since the counter example will have pointed the way towards a new problem and ultimately towards new knowledge. This explains why so many of the important discoveries in science have been "accidental". Ultimately, science is a problem solving activity, and progress in knowledge is made by the identification and solving of new problems.

Medawar (1969) has pointed out that viewing science as a hypothetico-deductive process provides a convincing explanation for scientific error—the fact that research so often goes wrong. Medawar (a winner of the Nobel Prize) estimated that four fifths of his time as a working scientist had probably been wasted on dead ends and mistakes. How can this be explained if scientific knowledge is based on reliable observations and a logical process of induction? On the other hand, the hypothetico-deductive process, with its emphasis on the conjectural nature of science and the importance of informed but bold guesses, shows us immediately how researchers may be led into error. Additionally, the hypothetico-deductive model shows how we can learn from our mistakes, and helps us to understand how creativity, intuition, luck and serendipity play such an important role in the research process.

> There is no logical path leading to [the highly universal laws of science]. They can only be reached by intuition, based upon something like an intellectual love of the objects of experience.
>
> Albert Einstein (1879–1955).

The philosopher Karl Popper claimed that this view of the way science makes progress provides a solution to Hume's problem. Simply put, *science does not use induction* and, indeed, the process of induction is a myth. Induction provides neither an explanation for the way in which scientists arrive at their theories, nor a justification for the truth of scientific knowledge. There can be no appeal to the authority of direct observation. Indeed, as we have seen, even direct observations with our senses may be mistaken. Our senses embody theoretically based structures and we interpret sense data with the preconceptions of our experience.

Furthermore, Popper has pointed out that the inductive view leads us to adopt a mistaken attitude towards the justification of scientific knowledge. Whether we believe that our knowledge is based on careful observations, or anything else, it is a mistake to seek an answer to the question "What are the best sources of knowledge—the most reliable ones, those which will not lead us into error, and those to which we can and must turn, in case of doubt, as a last court of appeal?"

(Popper 1974). Popper asserted that *no such ideal sources of knowledge exist.* Rather we should assume that *all* "sources" are liable to lead us into error at times. We should replace the question about the sources of our knowledge by the entirely different question: *"How can we hope to detect and eliminate error"*. The hypothetico-deductive model does not provide us with a royal road to guaranteed truth any more than the inductive model purported to do. However, it does allow us to distinguish between competing hypotheses. The processes of criticism and severe testing allow us to identify and reduce error so that we can make progress towards the truth, even though absolute scientific truth is ultimately unattainable.

We are thereby liberated into a world in which all scientific knowledge is conjectural and provisional. But this does not mean that our knowledge is arbitrary. On the contrary, through criticism and experimental tests we have the means to establish if we have made a mistake, and whether we are in error. And we can *learn* from our mistakes because, when our hypotheses fail, we can modify them or replace them with alternative hypotheses. Even if these are successful, we know that they, in their turn, may succumb to criticism. In this way, we make progress towards the truth, and this explains the success of science. It is not due to it being grounded in some secure, unassailable source of knowledge. It is because scientific theories have been subjected to a ferocious program of error elimination by criticism and severe test, and the rigor and integrity that is brought to bear on the research process.

Practical Points

- All scientific knowledge is provisional. Cultivate a skeptical but respectful attitude towards all scientific claims.
- Recognize the opportunities that arise from the provisional nature of scientific knowledge. Nothing is written in stone; even the most apparently unchallengeable theory may be mistaken.
- Cultivate the highest critical standards towards your own work. Develop your knowledge of experimental design and statistical analysis so that you can identify and eliminate error as efficiently and rigorously as possible. Replicate your own work, and provide enough information in reports and scientific papers for others to replicate it independently.
- Test rigorously, but do not abandon hypotheses too easily or too quickly. The mistaken rejection of a hypothesis may be the result of an insufficiently critical attitude towards the tests used.
- Learn from your mistakes. A negative or unexpected result may result from errors in experimental design or techniques. Such a mistake will provide an opportunity to improve your experiment or make it more robust. If your experiment is sound, then a negative result may shed light on errors, or assumptions in your thinking, that may direct you towards an improvement in your hypothesis.

- Recognize that the systems you use for testing of hypotheses may themselves be flawed. Despite your best efforts there may be errors in the quality of your thinking, in your experimental assumptions, or in your test instruments. Be humble—there are too many examples of ideas that remained part of the scientific orthodoxy until they were well past their sell-by date.

References

Magee B (1985) Popper. Fontana, London, p 24 (This is a first rate and accessible introduction to the thought of Karl Popper. It provides insight into how the rigor of philosophical thought in an apparently limited area can generate a coherent worldview.)

Medawar P (1969) Induction and intuition in scientific thought. American Philosophical Society, Philadelphia, p 24

Popper, KR (1974) Conjectures and refutations. Routledge and Kegan Paul, London, p 25 (The sections XIV–XVII, pp 24–30 at the end of the first chapter in this book provide an enlightening account of Popper's view of the sources of knowledge and ignorance.)

Chapter 10
Models: Useful but Not True

Abstract Our theories are, to some extent, models of reality. Like all models, a theory or hypothesis will tell us a truth about an aspect of reality, but it will not be the whole truth. All models are approximations of reality. Understand the limitations and assumptions of any theoretical model that you are using.

> ...all models are approximations. Essentially, all models are wrong, but some are useful. However, the approximate nature of the model must always be borne in mind...
>
> George Edward Pelham Box (Box and Draper 1987).

Our theories and hypotheses are representations of the real world. Through the twists, turns and dead-ends of research, these representations become progressively more accurate and precise. However, as scientists, we accept the provisional nature of our knowledge and that our understanding must remain forever incomplete. We grope our way towards a truth that is ultimately unknowable.

There is a sense in which all theories and hypotheses are simplifications and approximations of the real world in way that is similar to the manner in which a street map is a simplification and an approximate representation of the streets it portrays. Real streets have buildings that vary in height and appearance; they will have sidewalks, and will vary in width; they will have lights, traffic signs and signals, and dumpsters. The map does not provide any information about the height and appearance of buildings, and the streets may appear wider on the map than they are in reality. Nevertheless, we will expect the map to provide accurate information about the names of streets and their spatial relationships to one another. We accept the conventional relationship between maps and streets, and we agree that the map has practical utility despite its limitations. The map does tell a truth about the streets it represents, even if that truth is not the whole truth (Giere 1999).

We can call a map a "model" of the real world that it represents. In a similar fashion a plastic model of Columbus' Santa Maria is a representation of the real

P. Truran, *Practical Applications of the Philosophy of Science*,
SpringerBriefs in Philosophy, DOI: 10.1007/978-3-319-00452-5_10,
© The Author(s) 2013

ship that it portrays. We recognize it as a model of the real thing even though it cannot float, and has no functioning sails or rigging. On the other hand a detailed working model of a sailboat may be an extremely accurate representation of the full size boat, with differences only visible to an expert eye. But it is still a model. All our theories and hypotheses about the world can be considered to be models. Like models, our theories, and even our scientific laws, are approximations or representations of reality. They tell us the truth—but not the whole truth—about the real world. The simple theory or model of the relationship between the length of a spring and the weight applied to it (Hooke's Law) is approximate. It breaks down at its limits and does not take into account other variables, such as the temperature or the material used, that may affect the relationship.

Models will vary in the degree and the accuracy with which they represent reality. Theories such as quantum electrodynamics (the study of how electrons and photons interact) provide representations of reality that allow highly accurate predictions to be made. Although the predictions of quantum electrodynamics regarding the scattering of photons and electrons are accurate to ten decimal places (Feynman 2006), this powerful theory may still be regarded as a model of reality. By contrast, the theoretical systems used to model weather systems have a much lower level of accuracy and predictive power. This, of course, does not detract from their evident practical utility.

Even our primary senses, our sight and hearing, can be regarded as providing us with models of reality. The blue of the sky exists in our minds by virtue of the visual cortex's representation of the interactions between electromagnetic radiation of a certain wavelength and the cells of our retinas. There is no "blue" out there. Similarly, there is no sound except in our heads. Our ears and brains allow us to construct models (sound sensations) of the oscillations of pressure that are transmitted through the air or water that surrounds us. Our diagrams and equations describing sound waves are themselves models that allow us to better understand the phenomenon, and to make predictions that can be tested against reality.

> The sciences do not try to explain, they hardly even try to interpret, they mainly make models. By a model is meant a mathematical construct which, with the addition of certain verbal interpretations, describes observed phenomena. The justification of such a mathematical construct is solely and precisely that it is expected to work—that is, correctly to describe phenomena from a reasonably wide area.
>
> John von Neumann (1903–1957) (2000).

So why don't we just substitute the word model for theory—or vice versa? We don't usually use the term when discussing theories of propagation of sound or light, or when talking about genetics or evolution. We could do, but the term "model" is more often used by researchers when they need a way of finding new theoretical representations of the world that are worth testing against reality. Such models are constructions that allow researchers to play with different variables and constants, to explore the relationships between them and to explore the effects of making different theoretical assumptions. The model in this sense becomes tool for discovery; it has a "heuristic" function.

Perhaps one of the most famous examples of a model being used for a heuristic purpose was that used by Watson and Crick to work out the structure of DNA. They used flat steel plates that had been cut to represent the shapes of the purine and pyrimidine molecules that form the key constituents of the DNA macromolecule. By manipulating the arrangements of these plates in three-dimensional models they were able to identify different possible configurations, and compare the predicted X-ray crystallographic features with the actual data obtained by Wilkinson and Franklin. They explored the possibility that DNA might have a three-stranded structure, and were informed by Rosalind Franklin that the numbers of attached water molecules allowed by this structure were not consistent with previously known data. The double stranded structure, on the other hand, allowed the requisite amount of water to be accommodated within the molecular structure and was consistent with the X-ray crystallographic data. The manipulation of these relatively simple molecular models enabled Watson and Crick to visualize and identify the geometrical relationships between purines and pyrimidines, and ultimately to understand the mechanism for replication of DNA. The model bore a relationship to the real DNA molecule that was sufficiently true in geometrical and spatial terms to generate predictions that were accurate enough be tested against experimental data. Of course most scientific models are not made of steel, cardboard and paper clips. They will most frequently be in the form of diagrams, flow charts, mathematical formulae or computer algorithms.

The use of models is also well established in areas of research where the subjects of interest cannot be studied directly, either for reasons of accessibility, cost or for ethical reasons. For example, when medical researchers are investigating if a food additive causes cancer they will use animals rather than human subjects for their experiments. Rats are typically used because they are easy to rear in the laboratory and their physiology is known to be similar to that of humans in many important respects. Experimental results obtained in rats are, with important qualifications, considered to be reliable predictors of those likely to be obtained in humans and, accordingly, rats provide good models of human physiology. Of course, animal models are known to have limitations, and their use is subject to important ethical considerations. Nevertheless, in the case of food additives or pharmaceuticals, the demonstration of ill effects in animal models will be sufficient to provide a case against their use in humans. However, if new pharmaceuticals are to be licensed for humans, then their successful use in animal models will only provide the first step in the building of a convincing case for their clinical use. The data derived from animal models will be an essential part of assessing potential risks prior to embarking on studies with human subjects.

It is in the investigation of very complex phenomena that models come into their own. Consider the problem of the researcher investigating global warming. The hypothesis that global warming is taking place may be tested by examination of current and historical records of atmospheric temperature. However, the investigation of the potential effects of global warming involves large numbers of variables—increasing carbon dioxide levels, greenhouse gas emissions from cattle, cyclic variations in solar output, the changes in major ocean currents. None of these variables are amenable to manipulation in designed experiments, and so researchers will use mathematical models based on theories of thermodynamics, fluid dynamics and heat transfer. They will include models of atmospheric and oceanic phenomena, ice cover and biological processes. These models of climate change allow researchers to make and test predictions of the effects of global warming based on different scenarios.

Such a model becomes a useful means for manipulating the network of theoretical structures that represent our attempts to understand the phenomena of global warming and climate change, and becomes a device for generating new possibilities for testing. Different variables and constants can be introduced, and the mathematical relationships between them can be manipulated. The researcher generates new configurations of the model, new hypotheses, by using computer simulations that provide novel output and predictions. These new hypotheses may be assessed for plausibility and their predictions tested for correspondence with the real world. Therein, of course, resides the difficulty with the models of global warming and climate change. The predictions derived from computer models are difficult to test, and it may be decades or centuries before sufficient observations can be made to refute or confirm the causal relationship between human impact on the atmosphere and climate change. Accordingly, although computer models provide an essential mechanism for investigating climate change, they remain the subject of considerable controversy, and their power to influence economic and energy policy is accordingly limited. On the other hand, the use of complex models for weather forecasting is well established and, for short-term forecasts at least, has considerable credibility. We are all willing to plan our activities on weather forecasts, based as they are on limited predictions that are subject to day-by-day testing. We are also aware of the limitations of longer-term predictions and we are accordingly cautious in our use of them.

The use of modeling with computer simulations allows researchers to investigate phenomena that are particularly challenging because of their complexity or inaccessibility. This approach is extensively used in cosmological studies (for example, the formation of stars and galaxies), high energy and solar physics, economics or in the investigation of the processes of evolution. Computer simulations allow researchers to explore relationships between variables that might otherwise be too complicated to solve. However, there are dangers in their use. They may tempt researchers to develop increasingly complex models that, although they may give results that are empirically useful (such as in the case of the modeling of weather systems), may not be easy to relate to comprehensible theoretical mechanisms.

As models become more complex it is necessary to establish systems for their "verification" and "validation". Note that these terms are used widely in the context of engineering or software development. Verification of a model involves checking to establish that the mathematics and programming associated with the model are correct, and that the assumptions and simplifications are appropriate.

This is a technical job that may be beyond the capabilities of the researcher and will be done by the professionals who built the model. On the other hand, *validation* of a model will establish that it is consistent with its intended purpose. The model will need to be tested in ways that are analogous to the testing of any hypothesis. If testing shows that the model does not provide predictions that are in accordance with known knowledge then it may need to be revised or abandoned as invalid or "falsified". Of course, as with any hypothesis, confirmation of predictions does not constitute complete validation, any more than successful testing of a hypothesis will mean that it is true. Rather, confidence in the validity of a model will grow as it is verified, and critically tested and applied to the real world.

The notion that our understanding of the real world is based on models raises the question of whether science tells us anything about the real world or the nature of reality itself. This question becomes more pertinent when we consider that seemingly incompatible models may be used to make predictions about the same phenomenon. For example, the propagation of light was historically explained using models based upon views of light as either particles or waves. For each model we may believe that its predictive power is an indication of its being at least approximately true. But if both models are successful in making predictions, and yet mutually inconsistent, how can they both be true? Let us consider a simple illustration. Two observers are looking at a physical object. One may report seeing a circular disc, and the other may report seeing a rectangle. Both will be correct, but one will be looking at the object (a cylindrical can) from above and the other will be observing from the side. The two models represent different aspects of the same reality. In a similar fashion all models are, by their nature, limited representations of reality. They are created by us, and subjected to criticism and empirical test. We can elect to live with apparently contradictory models or we can attempt to integrate them; or we can derive a better model that accounts for the more limited models. In the simple case of the two models of the cylindrical can, we are able to develop a single three-dimensional model that encompasses the truths provided by each of the two-dimensional models. In the case of the propagation of light, we now understand that, in the quantum mechanical view of reality, all particles exhibit wave–particle duality—they have wave-like and particle-like properties. Classical (pre-1900) concepts like "particle" and "wave" are inadequate to fully describe the behavior of quantum-scale objects.

It is relatively easy to see that models, whether they are models of the weather, or models of stellar evolution, are human constructions. Our theories, our hypotheses, even our "facts", are also human constructions. We accept that they tell us truths about the world that are to some degree approximate, and that will depend upon the assumptions made and on the variables chosen for the model. Forecasters on the Weather Channel will openly illustrate the effects of using different models on their predictions, and it is evident that we are becoming increasingly comfortable with the notion that our view of the world may be dependent upon the models we choose. This same fluidity of perspective may be extended to all our knowledge.

However, few researchers question that the world that they study truly exists, and that they can know about it directly through their senses. Furthermore, they generally believe in the reality of entities that we can observe directly and also those theoretical entities that cannot be directly observed, such as atoms or electrons or viruses. Philosophers of science call this view *scientific realism*, and it is generally accepted to be the case for observable entities. Scientific realists point to the progress of science, and its predictive successes, as evidence for the existence of non-observable as well observable entities. One recent alternative to scientific realism is *constructive empiricism* which has been proposed by Bas van Frassen (1980). Constructive empiricism accepts that science aims to establish the truth about the observable world, but claims that it cannot aim to tell the truth about non-observable entities. An empirically adequate theory or model about non-observables will not claim to be true, but it will allow things to be said about the observable things and events in the world that *are* true. Thus, we now know that Newton's theory of gravitation is not true (having been supplanted by Einstein's theory), but it remains empirically adequate. This may be a useful perspective for the researcher to adopt since it may be used to justify the use of models and theories which may incorporate entities, concepts or terms which may not be claimed to be true, but rather contribute to the achievement of empirical adequacy. Such theories or models, in addition to being practically useful, may provide results that eventually lead to new interpretations of current knowledge (Wimsatt 2007) and to better theories.

Practical Points

- Recognize that all our theories are, to some extent, models of reality. Like all models, a theory or hypothesis will tell us a truth about an aspect of reality, but it will not be the whole truth. All models are approximations of reality.
- When a theory becomes too complicated to understand or manipulate, a simplified model, or a computer simulation, may be used to establish if the theory is on the right track. The model or simulation becomes a powerful means for identifying and assessing new possibilities for further investigation and empirical test.
- It may be necessary to create a model that takes a totally different perspective in order to improve upon currently accepted models. Creating such a model will require true boldness of vision.
- Understand the limitations and assumptions of any theoretical model that you are using. Such limitations should be made explicit when you set out to use the model.
- Carefully consider the issues around the verification and validation of models that you either use or develop. Verification of the model will address its internal consistency and logical coherence; validation will provide evidence that the model provides reliable information and predictions about the real world. The approach to be taken will depend on the type of model (physical, visual, mathematical, conceptual) and the research fields in which they are being employed.

References

Box GEP, Draper NR (1987) Empirical model-building and response surfaces. John Wiley & Sons, New York

Feynman R (2006) QED: the strange story of light and matter. Princeton Science Library, Princeton, NJ p 7

Giere RN (1999) Science without laws. University of Chicago Press, Chicago, p.81

Van Fraassen B (1980) The scientific image. Oxford University Press, Oxford

Von Neumann J (2000) Method in the physical sciences. In Bródy F, Vámos, T (eds) The neumann compendium. p 628

Wimsatt WC (2007) Re-engineering philosophy for limited beings: piecewise approximations to reality. Harvard University Press, Cambridge

Chapter 11
Just Be ...... Cause

Abstract It may be extraordinarily difficult to unequivocally demonstrate a causal relationship. Presenting clear evidence of causality will be of particular importance for the researcher when it becomes necessary to convince others of a relationship that will be the basis for decision-making and social policy issues.

> All reasonings concerning matters of fact seem to be founded on the relation of cause and effect.
>
> David Hume (1711–1776)
>
> I would rather discover one causal relation than be King of Persia.
>
> Democritus (430–380 BC)

As human beings, we seem to have an extraordinary psychological need to identify causal relationships. It is intuitively obvious that this need originates from our desire to control and manipulate the environment around us. From the time that a baby starts to move and then establish control of its fingers or toes, to shake a rattle to make a pleasing noise, or to smile to attract the attention of its parents, we see the phenomenon of control through the identification of causal relationships. We are defined as a species by our ability to identify such relationships and to use this knowledge to devise tools with which to change our world. The establishment of a causal relationship provides a key element in the identification of a solution to a problem.

This need to identify causal relationships may be an important aspect of our success as a species, but it may also lead us into error. We constantly look for patterns of causality, and the most obvious are those patterns of events that regularly happen in time sequence. I put my hand in the flame and I feel the burning. I press the brake and the car stops. Such relationships are seen so frequently that we have a tendency to see them everywhere. I rub a patent medicine into my back—and my backache gets better. A comet appears in the sky—and an important event occurs. We have a psychological investment in believing in causal relationships because they provide us with the illusion of control, and we may believe that a causal relationship exists on the basis of the most flimsy evidence. A few anecdotes that seem to provide evidence of a cure for cancer may be sufficient to persuade sufferers to embark on expensive programs of treatment that offer hope and promises but nothing more. We might think that the application of reasoned thinking and testing by experiment would provide a clear route to establishing causality. However, from both philosophical and scientific perspectives, it may be extraordinarily difficult to provide convincing evidence of a causal relationship.

The relationship between cause and effect has been one of the principal concerns of philosophers ever since time of Aristotle, and it remains an active area of philosophical inquiry and debate. Philosophers have argued energetically about the nature of causality. However, by identifying criteria for establishing causality, they have helped scientists to convincingly show clear links between phenomena under study and their proposed causes. In the 18th century, the British philosopher David Hume provided an account of causality that has been widely used ever since. He proposed a list of ways in which we might establish a cause and effect relationship between two things. This list included the following criteria:

1. Cause and effect must be close to one another in space and time.
2. The cause must be prior to the effect.
3. There must be a constant relationship between the cause and effect.

Hume also claimed that the same cause always produces the same effect, and that where several different causes produce the same effect, there must be something common between them.

Now, consider Hume's third criterion. It may be appear to be relatively easy to demonstrate a constant and repeatable relationship between two sorts of phenomena. Every time put my foot on the accelerator pedal, the car goes faster; every time I take a couple of aspirin, I get relief from my headache. This sort of constant relationship (correlation) seems to provide powerful evidence of a causal relationship. However, the uncritical establishment of a correlation may lead us to an erroneous inference of causality. The relief I get from the aspirin may be due to a placebo effect rather than the effect of the drug itself. Similarly, the improvement in a back injury that I apparently get from a course of physical therapy may be due to the natural healing of the body rather than the therapy. It is a well-known principle that correlation does not necessarily imply causation. Nevertheless, as Tufte (2006) stated (with some humor), "Correlation is not causation but it sure is a hint."

Measurements of global temperatures and atmospheric carbon dioxide levels have been made over the last two hundred years. As carbon dioxide levels have been rising so too have global temperatures, and these two variables can be said to be *positively* correlated—as one variable increases, so does the other. On the other hand, there may be situations where the association between two variables is such that, as one increases, the other *decreases* in a systematic manner. In this case the two variables are said to be *negatively* correlated. For example, the incidence of infectious disease decreases as the rate of immunization increases. When we have evidence of a correlation (either positive or negative) between two variables we may jump to the conclusion that one phenomenon is the cause of the other. Accordingly, as a result of demonstrated correlations, it is proposed that the rise in carbon dioxide is the *cause* of global warming, and that programs of mass immunization have been the *cause* of reductions in the incidence of many previously endemic diseases.

Although the evidence for a correlation between atmospheric carbon dioxide and global temperatures is overwhelming, the case for a causal relationship is by no means incontrovertible. Global warming may have other causes, such as changes in solar activity. The debate about whether global warming is caused by man-made changes in carbon dioxide concentrations continues to affect the ability of policy makers to agree on the appropriate counter measures. Similarly, although there may be a clear relationship between the incidence of tuberculosis in western societies and immunization polices, it is likely that other causal factors, such as improved housing and changes in social habits, are also important. Researchers whose findings may have significant implications for medical, social or economic policy will find it necessary to pay particularly close attention to issues of multiple causality if they are to provide convincing arguments in the face of established interests and beliefs.

Providing adequate evidence of causality may be challenging when a particular phenomenon has multiple causes or a causal system incorporating a network of indirect causes, probabilistic causes, feedback loops and interactions. This is especially common in climatic systems, biological systems, ecosystems, and social and economic systems. Researchers also need to be aware of the notion of the *common cause*. Consider two geysers in a volcanic area, separated by hundreds of yards, that erupt periodically and simultaneously. The events are correlated, but we would not suggest that one is causing the other. Rather, we would propose a single common cause, the regular heating of a subterranean reservoir of water that supplies both geysers. Similarly, red hair may be correlated with blue eyes, but one is not the cause of the other. These correlated characteristics stem from an underlying genetic cause that is common to both conditions. The issue of common causes may be important when trying to establish direct causal relationships, such as those between the incidence of infectious diseases and patterns of immunization, where increases in living standards may provide an apparent common cause for both phenomena. In the case of the proposition that cigarette smoking causes lung cancer, we need to rule out the possible common cause that smokers are more likely to live in urban areas with higher levels of air pollution.

In the physical and biological sciences, and in engineering, it may be relatively easy to establish causation. Let us take a relatively simple example like the investigation of the relationship between a drug and its effect—a dose-response relationship. Studies done on a model organism such as the rat will allow the causal variable (the dose) to be manipulated to change the response effect. In carefully controlled experiments, the researcher can eliminate the effects due to other potentially interfering variables such as diet, sex and activity. Typically, when the drug dose is deliberately increased, the response is also increased; when the dose is reduced, the response is reduced. All of Hume's criteria for demonstrating causality are satisfied, but it is the ability to *manipulate* the dose and its associated response that provides the most persuasive evidence of a causal effect. The causal relationship seen in such dose-response effects is similar to that seen in engineering, where the setting of a control variable will reproducibly bring about a change in a response variable. Dial in an increase in temperature and the pressure increases; turn it down and the pressure decreases. Such causal relationships, where it is possible to manipulate a cause variable to demonstrate a reproducible effect, are ubiquitous within the physical and biological sciences. Of course, there are some areas of physics, such as quantum mechanics, where causality seems to breaks down and physicists no longer find it useful or appropriate to provide explanations in terms of causal language.

In situations where it is not so easy to manipulate causes to elicit predicted effects, it may be much more difficult to establish causality. The famous epidemiological study conducted in 1954 by Doll and Bradford Hill provides an exemplary account of the steps that a researcher may have to take to provide evidence of a causal relationship when manipulation is not possible (Doll 2000). A powerful case was necessary to show a causal relationship between smoking and lung cancer, and thereby convince legislators of the need to change smoking habits. Their classic study reported on the rates of cancer and other diseases in a sample of doctors of medicine, and demonstrated that they were correlated with their smoking habits. However, for practical and ethical reasons, studies involving manipulation were not possible in humans. The conclusions derived from these epidemiological studies were therefore supported by laboratory studies that involved manipulation of cause (dose) variables and measurement of effects in animal subjects. However, in the struggle to get their work accepted as showing a causal relationship, Doll and Hill had to carefully consider how they could convincingly demonstrate causality.

In 1965 Bradford Hill published a list of criteria that stipulated guidelines for providing evidence of a causal relationship in epidemiological studies (Bradford Hill 1965). These criteria, which were clearly based on Hume's analysis of causality, have been widely accepted as being applicable in the health sciences in general, and they may be applicable to complex situations such as those found in the social sciences. They provide a checklist that, while not uncontroversial, is particularly useful in making decisions when the supporting evidence is strong but not conclusive (ISIS 2013).

Bradford Hill's criteria are as follows:

1. Temporal Relationship. Cause must be prior to effect. The cause of a disease must of necessity always precede the occurrence of the disease. Smoking precedes the vast majority of lung cancer cases.
2. Strength of association. The stronger the statistical association between two variables the more likely it is that the relationship is causal. In the case of Doll and Hill's study, those smoking 25 or more cigarettes per day were 32 times more likely to die of lung cancer than non-smokers.
3. Consistency. The relationship between two variables is considered to be consistent when the results may be replicated using different methods. In the case of smoking and lung cancer many studies were done before the causal relationship between the two was accepted.

4. Coherence. In order to support a case for causality, there must be a logical or theoretical mechanism that would explain the relationship. For example, although there may be a correlation between the rise of religious attendance through the 1990s and use of illegal drugs, there would appear to be no logical basis for concluding that religious attendance causes drug use. In the case of smoking there is a biological basis for the disease.
5. A demonstrated dose-response relationship. In the case of smoking, Doll and Bradford Hill showed that an increase in consumption of cigarettes was associated with an increased risk of incurring the disease.
6. Experimental evidence. Studies carried out on mice and dogs provided powerful evidence of carcinogenic agents in tobacco smoke.

Hill's criteria also include considerations of plausibility, specificity, and analogy (the consideration of alternate explanations). In the case of smoking and lung cancer, the evidence provided an overwhelming case for a causal relationship. In 1964, based on the epidemiological evidence, the Surgeon General issued a report linking cigarette smoking to death, cancer and, in particular, lung cancer. Inevitably these conclusions were challenged by the tobacco industry who, with the backing of the eminent statistician Sir Ronald Fisher, claimed that the apparent correlation could be explained by a common cause that was, as yet, undiscovered. Fisher postulated that there might be a genetic configuration, common to smokers, that simultaneously caused both lung cancer and a craving for nicotine. Ultimately, the evidence from laboratory studies, and the impact of socially engineered reductions in the incidence of smoking in the population, resulted in the widespread acceptance of a causal relationship.

Now of course it is also true that smoking does not inevitably lead to the development of lung cancer. We are all familiar with the counter example of the grandmother who lived to a ripe old age despite smoking a pack a day for most of her life. Smoking increases the *probability* of developing lung cancer; it does not inevitably lead to the disease. The relationship between smoking and cancer is therefore not a strictly physically *deterministic* one in which causes are inevitably followed by their effects. Indeed it is now generally accepted that we do not live in a deterministic universe. The Newtonian notion of a universe in which celestial bodies move with entirely predictable, clock-like motions has been supplanted by the development of a relativistic and quantum mechanical view. Atomic clocks can be used to detect minor fluctuations in the orbit of the earth that are not deterministically predicable. We now know that at a fundamental level the world is best understood in probabilistic terms, and that conventional notions of causality may no longer apply.

> It is that men's conceptions of a cause are in different stages of scientific culture entirely different and inconsistent. The great principle of causation which, we are told, it is absolutely impossible not to believe, has been one proposition at one period in history and an entirely disparate one at another is still a third one for the modern physicist. The only thing about it which has stood... is the name of it.
>
> Charles Sanders Peirce (1839–1914)

Although arguments about the nature of causality can be absorbing, they may seem academic. However, most of us do not want to persist in the illusion that our actions are useful and effectual when they are, in fact, not so. In the world of experimentally based research the establishment of causality may not be so problematic because interventionist strategies are commonplace. However, where intervention is not possible, the researcher may have to pay particular attention to the arguments centering on causality. This will be of importance for the researcher when it becomes necessary to convince others of a causal relationship that will be the basis for determining action. This will be the case for the physical, biological and medical sciences, but it will be especially important in the social and behavioral sciences and in economics, where experimental manipulation of variables is more difficult. In these areas, an understanding of the nature of causality and, in particular, the application of the Bradford-Hill rules, may be helpful to the researcher who wants to establish evidence to support wider decision making and social policy issues.

Practical Points

- It may be extraordinarily difficult to unequivocally demonstrate a causal relationship.
- The case for demonstrating causality will often be circumstantial. The rigor with which a causal case is presented will depend to some degree upon the consequences of demonstrating a causal relationship.
- The concept of causality employed in a field of research, and the degree to which it is deterministic or probabilistic, will depend upon the nature of that field and upon the specific phenomenon under investigation. It is important to choose appropriate criteria for demonstrating a causal relationship.
- Do not immediately dismiss an association that does not seem plausible or probable. This may be a new relationship that may turn out to be true.
- Problem solving may require the application of techniques for establishing causal relationships such as system diagrams or cause and effect (fishbone) diagrams. The mapping of causal relationships is a powerful starting point for understanding the nature of a problem and the generation of hypotheses or models.
- Complex causal systems will be particularly amenable to being modeled so as to identify and clarify the relationships between causal variables. It may be tempting, for practical reasons, decide to look at one causal variable at a time while holding the others constant. However, this may be misleading since it will not provide information about the patterns of *interactions* between variables in the causal system. In such cases more sophisticated experimental designs, such as factorial or partial factorial designs may be indicated.

References

Bradford Hill A (1965) The environment and disease: association or causation? Proc R Soc Med 58:295–300

Doll R (2000) Smoking and lung cancer. Am J Respir Crit Care Med 162:4–6

ISIS Report 03/02/10. The Bradford hill criteria applied to climate change and GMOs. Institute of science in society. http://www.i-sis.org.uk/TheBradfordHillCriteria.php. Accessed 1 Mar 2013

Tufte ER (2006) The cognitive style of powerpoint: pitching out corrupts within. Graphics Press, Cheshire

Chapter 12
A Question of Trust

Abstract The high profile cases of scientific misconduct which have surfaced in recent years have highlighted the need for guidelines to help researchers avoid drifting over the line into fraud, and to ensure that honest researchers are in a position to effectively guard themselves against such accusations. Ultimately, the integrity and credibility of science depends upon the intellectual honesty of its researchers.

> Intellectual honesty does not consist in trying to entrench, or establish one's position by proving (it) ...... intellectual honesty consists rather in specifying precisely the conditions under which one is willing to give up one's position.
>
> Imre Lakatos (1970)

In 1998, *The Lancet*, a respected British medical journal, published a paper that claimed to identify links between a new form of bowel disease, autism and the MMR vaccine used to vaccinate children against the three diseases Measles, Mumps and Rubella. The paper was authored by Andrew Wakefield, a surgeon carrying out research at the Royal Free Hospital in London, England, and twelve other doctors. It reported that, in eight out of twelve children showing symptoms of autistic behavior, the symptoms had started within two weeks of the children being vaccinated. The paper rapidly became the center of an on-going controversy about the role of the MMR vaccine in the onset of autism in children. Although a study carried out in Japan showed no causal link, Wakefield's paper, together with popular concern stoked by uncritical media coverage, resulted in calls for the withdrawal of the combined MMR vaccine and its replacement by separate vaccinations for each of the three diseases. In the UK public confidence in the combined vaccine was damaged and vaccination rates plummeted. In some parts of London vaccination rates fell below the levels that could be expected to prevent an epidemic.

As the vaccination rates in the UK and Ireland fell, the incidences of measles and mumps increased. Mumps had been a rare disease in the UK for many years

P. Truran, *Practical Applications of the Philosophy of Science*,
SpringerBriefs in Philosophy, DOI: 10.1007/978-3-319-00452-5_12,
© The Author(s) 2013

prior to the controversy about the MMR vaccine. After 1999 the incidence of mumps began to increase until by 2005 there was a full-blown epidemic. In 2008 measles was declared to be endemic in the UK—the disease was sustained within the population as a result of the low vaccination rates in the previous ten years. In 2006 there was the first death from measles in the UK since 1992. In Ireland three children died in the Irish measles outbreak of 2001 that occurred as a result of the reduction in MMR vaccination rates following the controversy.

In 2004 Brian Deer, an investigative journalist working for The Sunday Times of London, reported that Wakefield had received substantial payments from law-yers who were pursuing litigation on behalf of parents who claimed that the MMR vaccine had damaged their children. Wakefield had not reported this con-flict of interest to his co-authors, nor to the authorities at the Royal Free Hospital. Furthermore, Deer's investigations revealed that, for all twelve of the children in the study reported in *The Lancet*, the data on the diagnoses of the children and the dates of onset of symptoms had been altered to fit the conclusions of the report. *The results had been faked.* It was also established that the polymerase chain reaction studies done to detect the presence of measles were fatally flawed and could not possibly have obtained the results that were reported. The study had required invasive and distressing tests of the child subjects including anesthesia, ileo-colonoscopies, lumbar punctures, brain scans, EEGs, radioactive drinks and x-rays. Crucially, Wakefield had falsely reported in the paper that he had received approval of the study from the Royal Free's ethics committee. Deer also revealed in 2011 that Wakefield had hoped to profit from litigation arising from the vac-cination scare by launching a venture to market diagnostic kits and alternative vaccines.

As a result of Deer's investigations (Deer 2011) the anti-MMR campaign in the UK collapsed and vaccination rates rebounded. In February 2010 the UK General Medical Council reported an inquiry that found Wakefield guilty of dishonesty and abuse of developmentally challenged children. The Lancet fully retracted the 1998 paper. In May 2010 Wakefield was struck off the medical register of the UK General Medical Council, and in 2011 the British Medical Journal (Godlee et al. 2011) declared that Wakefield's research had been "an elaborate fraud". The Royal Free medical school and *The Lancet*, which had both carried out investi-gations of Wakefield and found no cause for major concern, were accused of "institutional and editorial misconduct". The Wakefield affair represents a clas-sic case of research misconduct, and the consequences were widespread and hor-rendous. Children suffered unnecessary disease; some died. Pediatric patients were subjected to painful, distressing and inappropriate medical tests and proce-dures. Careers were destroyed and the reputations of institutions were damaged. Legitimate research was disrupted and the public status of science was affected.

Extreme cases such as those of Wakefield are uncommon but not rare. The more recent case of Scott Reuben, Professor of Anesthesiology and Pain Medicine at Baystate Medical Center in Springfield, Massachusetts, provides another illus-trative example of deliberate fraud. Research was carried out with funding that had been provided by organizations that had an interest in a particular outcome from the research program. Reuben fabricated much of the data in 21 clinical trial papers that supported claims of safety and effectiveness for painkillers and anti-depressants marketed by the pharmaceutical companies Pfizer, Merck and Wyeth. Reuben received substantial research grants from the corporations that benefited from his support of their products. In 2010 Reuben was convicted of fraud in a Federal Court and sentenced to six months in jail.

> Misconduct in research (for example, fabrication, falsification, and plagiarism) damages the scientific enterprise, is a misuse of public funds, and undermines the trust of citizens in science and in government.
> Best Practices for Ensuring Scientific Integrity and Preventing Misconduct
> OECD Global Science Forum (2007)

The financial temptations and potential conflicts of interest that may lead to fraud are pervasive. But there are other temptations. A successful career in scientific research is very much dependent upon building a good publication record. Researchers may be pressured to produce high-profile papers or to rapidly deliver results in order to establish priority. These pressures may push otherwise honest researchers to cut corners or to be insufficiently self-critical. And it may be relatively easy for researchers who do falsify data to escape detection. Much research is never replicated because such work is not considered to be worthy of professional kudos. Even if work cannot be replicated, it may be possible for the researchers to claim innocence of fraud because of the effects of unconsidered interfering factors. Considerations of the history and sociology of science, as well as the philosophy of science, can also provide useful perspectives on these issues.

The high profile cases of scientific misconduct that have surfaced in recent years have highlighted the need for guidelines to help researchers to avoid drifting over the line into fraud, and to ensure that honest researchers are able to effectively guard themselves against such accusations. It is important that budding researchers are familiar with the standards that are expected of them. The booklet "On Being a Scientist", produced by the National Academy of Sciences (Committee on the Conduct of Science. National Academy of Sciences 1989), provides an excellent analysis of the requirements for avoiding misconduct in research. Such misconduct is defined as "fabrication, falsification, or plagiarism in proposing, performing, or reviewing research, or in reporting research results". Fabrication is described as making up and reporting results. Falsification, on the other hand, is the changing or omitting of data so that the research record does not accurately report what was done (falsification in this sense is to be distinguished from the refutation of a hypothesis). Plagiarism is "the appropriation of another person's ideas, processes, results, or words without giving appropriate credit". It is important to note that "Research misconduct does not include honest error or honest differences of opinion".

Let us assume that those who do fabricate data have made a conscious and deliberate decision to deviate from the attitudes and ethics that underpin scientific research. However, in the areas of falsification and plagiarism it is possible for the otherwise honest researcher to unwittingly stray over the line into misconduct. The standards that are now expected with respect to the handling and manipulation of data are high. The researcher's best defense against any charge of falsification or fabrication of data is to be scrupulous with respect to the documentation of research. In the past such standards were not always met. A good example is provided by Robert Millikan's oil-drop experiments to determine e, the charge on the electron, and for which he was awarded the Nobel Prize in 1923. Millikan kept detailed records of his laboratory work, which involved difficult and delicate measurements of microscopic oil droplets that were manipulated to rise or fall within an electric field. When these records were examined as part of a study done in the 1970s (Goodstein 2000) it was discovered that, of the 175 data points recorded in Millikan's notebooks, only 58 were reported in his paper of 1913. Millikan reported that these measurements were not from a selected group, but were made on 60 consecutive days. His notes make it clear that data were actually selected to meet with his expected value for e, his theoretical preconceptions. The selection of data points allowed the value for e to be stated with greater precision. As a result of the 1970s study of the data, real concerns were voiced about Millikan having committed fraud in the reporting of his work. Although careful review of Millikan's laboratory notebooks and analyses of data exonerated him of any deliberate fraud, the case provides an instructive example of the standards that are now expected of researchers. We have seen that there were similar question marks about the selection of data in Eddington's famous eclipse experiment to test the predictions of Einstein's General Theory of Relativity.

The cases of Eddington and Millikan may give us cause to think that having theoretical preconceptions of the expected results *prior* to the implementation of an experiment may lead to the selection or massaging of data to meet with these preconceptions. However, we have previously seen that, before carrying out an experiment, it is important for the researcher to have a clear idea of the results that will confirm or refute the hypothesis under test. Such considerations will clarify how attitudes, prejudices and preconceptions may affect the outcome of research work. They will help the researcher to identify the appropriate experimental designs and statistical analyses necessary to reduce the effects of such attitudes on the test. These will include the use of blind or double blind studies, controls etc. This approach, when well documented by the researcher, is likely to *minimize* the possibility of charges of misconduct. Of course, the researcher may have a legitimate need to select or manipulate data. In such cases there are well-established procedures for identifying statistical outliers, and for eliminating observations that have been attributed to experimental error. Whatever the reasons for the elimination or manipulation of data, it is essential that these are transparent and acknowledged in the final report. Raw data, and the records of the data manipulation, must be available for review, preferably in the form of well managed and witnessed laboratory notebooks. Good experimental design must be matched by good laboratory practice. The failure of work to be replicated by other workers may lead to accusations of fraud when the actual problems may be poor laboratory technique and quality systems. Good technique is rooted in a solid understanding of the materials, equipment and techniques being used for research. Poor training, inadequate calibration of measurement devices and sloppy record keeping may also get the researcher into trouble if it leads to difficulties in replication.

> Science is at once the most questioning and ... skeptical of activities and also the most trusting. It is intensely skeptical about the possibility of error, but totally trusting about the possibility of fraud.
> Arnold Relman, former editor of the New England Journal of Medicine

Ultimately, the challenge of maintaining integrity as a researcher will be dependent on having an attitude that is consistent with scientific honesty. Research involves doing tests to allow choices between available hypotheses, with the researcher making the tests as severe as possible. Scientific honesty requires that the conditions for rejection of a hypothesis should be specified *in advance* of carrying out the tests recognizing that, because the process is fallible, we may make the wrong choice. We may eliminate a true hypothesis and accept a false one. This view of scientific honesty requires that the researcher be continually alert to the possibility that he may be mistaken. It requires that hypotheses be formulated as clearly and unambiguously as possible (in terms of a Null-Hypothesis) and that experimental designs be drafted rigorously to afford severe tests. Above all it requires the researcher to accept the uncomfortable fact that a favored hypothesis may have to be rejected, even if this may have personal, career or financial implications.

This is where the temptations to falsify data may be greatest. The researcher may have a gut feeling that he is right or that a hypothesis is so beguiling it just has to be true. Any data that get in the way of demonstrating the correctness of the hypotheses must be due to procedural errors or bad luck. The excitement of discovery may tempt him into crossing the line into misconduct. This is where it is so vital for the researcher to have peers and colleagues with whom he can frankly share his ideas, his enthusiasms and his results. Research is a communal activity and there are real dangers in working in isolation. Yet the researcher who has developed new, interesting ideas may feel that he is working in an environment where such ideas may be unfairly discounted or misappropriated. It is vital that he seeks out colleagues and advisors who can be trusted, and with whom he can discuss his work. In addition to the cultivation of an attitude that embraces self-criticism and procedural rigor, it is vital that the researcher embraces the need to be open about his research. Ultimately, despite the systems of checks and balances that exist within the system, the peer review procedures, and the options for replication of findings, the integrity and credibility of science will depend upon the intellectual honesty of its researchers.

Practical Points

- Accurately maintain primary records, the output from measuring devices, raw observations, photographs or scans, recordings, survey returns, interviews.
- Ensure that raw data, and the records of the data manipulation, are available for review, preferably in the form of well managed and witnessed laboratory notebooks.
- Use laboratory notebooks, which are bound, have numbered pages and are reviewed and signed on each page by a colleague or supervisor. Include raw data printouts (signed and dated), raw calculations, accurate descriptions of experimental procedures and statistical analyses.
- Never be careless about record keeping. Write up designs and techniques prior to doing an experiment. Record results and analyses immediately after doing the work.
- Match good experimental design with good laboratory practice. The failure of work to be replicated by other workers may lead to accusations of fraud, when the actual problems may be rooted in poor laboratory technique.
- Specify the conditions for rejection of a hypothesis *in advance* of carrying out experimental tests.
- Cultivate good working relationships with statisticians, and with those who are carrying out analytical or technical procedures on your behalf.
- Ensure that you are working in an environment where you can frankly discuss your ideas, your experimental designs, and your results with colleagues and advisors.
- Familiarize yourself with the guidelines for avoiding research misconduct laid out in "On Being a Scientist" produced by the National Academy of Sciences.

References

Committee on the Conduct of Science. National Academy of Sciences (1989) On being a scientist. Proc Nat Acad. Sci USA 86:9053 (An excellent guide to ethical principles for scientists, published as a downloadable book in 2009 by the US national academies of sciences. It's very accessible for graduate and undergraduate students, and includes case studies for discussion.)

Deer B (2011) Secrets of the MMR scare: how the case against the MMR vaccine was fixed. BMJ 342:c5347

OECD Global Science Forum (2007) Best practices for ensuring scientific integrity and preventing misconduct. http://www.oecd.org/dataoecd/37/17/40188303.pdf. Accessed 31 Aug 2011

Godlee F, Smith J, Marcovitch H (2011) Wakefield's article linking MMR vaccine and autism was fraudulent. BMJ 342:c7452

Goodstein D (2000) In defense of Robert Andrews Millikan. http://eands.caltech.edu/articles/Millikan Feature.pdf. Accessed 31 Aug 2011

Lakatos I (1970) Falsification and the methodology of scientific research programmes. In: Lakatos I, Musgrave A (eds) Criticism and the growth of knowledge, Cambridge: University Press, Cambridge, p 92

Chapter 13
Science, Non-Science and Pseudoscience

Abstract Science is under attack from those who believe that truth is relative and from those who reject the critical standards of science in favor of comforting pseudoscientific beliefs. Be prepared to defend science by understanding the philosophy of science so that you can marshal your arguments against pseudoscience.

> If only the ignorant and gullible were swayed by far-fetched claims, little else would be needed to explain the abundance of folly in modern society. But oddly enough, many people who are neither foolish nor ill educated cling to beliefs repudiated by science.
>
> Barry Beyerstein (2001)

We have seen that the project to establish a sharp line of demarcation between science and non-science (or metaphysics) is probably doomed to failure. It is now accepted that scientific researchers will try to formulate hypotheses that can be rigorously tested in such a way that there is a possibility that they may be falsified. However, falsification can always be avoided, and therefore science is better characterized by an *attitude* that embraces testability and the possibility of falsification, rather than trying to shore up hypotheses against any possibility of their being disproved.

In the domain of metaphysics, key statements and concepts do not claim to be falsifiable, even in principle. For example, theological discourse will be based around a concept of God which, it is agreed, cannot be tested or falsified by any means. Neither can the idea that it is wrong to treat animals cruelly. This does not mean that metaphysical statements are meaningless. We accept that the religious, ethical or philosophical systems of thought around which we organize our lives cannot be subject to empirical test and potential falsification. However, they can be subjected to rational analysis and criticism, just as scientific theories can. For example, the idea that it is unethical to use animals in medical research may be examined and debated rationally. We can invoke arguments based upon the perceived benefits of such research to human health, the scientific evidence of pain

P. Truran, *Practical Applications of the Philosophy of Science*,
SpringerBriefs in Philosophy, DOI: 10.1007/978-3-319-00452-5_13,
© The Author(s) 2013

felt by animals subjected to experiments, the protections put in place to ensure animal welfare and the options for alternatives to the use of laboratory animals. Such moral arguments are recognized as being valid and meaningful even though there can never be an appeal to empirically based data in order to decide the issues. Metaphysical questions cannot be decided by experiment.

Karl Popper showed how some metaphysical theories, empirically untestable when they are originally formulated, may incorporate truths that eventually become amenable to scientific testing. He cites the example of atomism which, as developed by the Greeks in the 5th century BCE, was not testable. The philosopher Democritus proposed that the material world was composed of tiny, indivisible particles that were the building blocks of all matter. He explained the sensations of hot and cold in terms of the different packings and scatterings of the atoms that comprised the objects. The atomists deduced that there must be different types of atoms to account for the variety of forms that we experience in nature. They also recognized that atoms exist in a void, and that the differences seen in the densities of different materials were due to the differing amounts of void in them. None of this was subject to empirical test, and this perceptive metaphysical theory was eventually rejected after the time of Aristotle. It was not revived in Europe until the 17th century. We now know that many of the conclusions that derived from Greek atomic theory were in accordance with a scientific conception of the atom that was accepted right up to the end of the 19th century, and which continues to have enormous explanatory power. This pre-scientific metaphysical concept therefore incorporated truths that were ultimately corroborated by empirical scientific testing.

> If, in some cataclysm, all of scientific knowledge were to be destroyed, and only one sentence passed on to the next generation of creatures, what statement would contain the most information in the fewest words? I believe it is the *atomic hypothesis*.... that *all things are made of atoms - little particles that that move around in perpetual motion, attracting each other when they are a little distance apart, but repelling upon being squeezed into one another.* In that one sentence, you will see, there is an *enormous* amount of information about the world, if just a little imagination and thinking are applied.
>
> Richard Feynman (1996)

So we may not be able to sharply distinguish between scientific and metaphysical ideas. The testing of scientific theories may be a matter of attitude rather than iron logic; metaphysical ideas may not be falsifiable or testable, but they may contain truths that may eventually be accepted as scientific. We generally manage to accept this kind of ambiguity, and in most cases the assignment of the labels "scientific" and "metaphysical" is uncontroversial. We recognize the empirically based authority of science for descriptions of the physical and biological domains; we are able to distinguish the metaphysically characterized areas of our lives that do not need to make any claims on the authority of science to establish their validity. We understand, despite the grey areas between the two domains, that science and metaphysics generally remain distinct. However, our critical antennae will generally go onto the alert when scientists start to apply scientific principles to issues of religion or ethics, or when metaphysicians invoke science to justify their arguments. It is in the latter situation that we need to be aware of the dangers of pseudoscience.

The term pseudoscience is applied to beliefs or practices that unjustifiably claim the status of science. No doubt we are familiar with the multitude of ideas or products, often with associated equipment and materials, that present the appearance of being based upon scientific principles. A clear warning sign used to be the appearance, in an advertisement, of an earnest, middle-aged "scientist" in a laboratory coat. This technique is used less commonly now, but there are still plenty of attempts to establish scientific credibility by using scientific terms such as vibration, quantum or energy. There may be associated pseudoscientific equipment: electromagnetic field (EMF) meters for detection of ghosts, the E-meter used in Scientology, the orgone accumulator used by Wilhelm Reich "to concentrate atmospheric orgone".

The enormous variety of pseudoscientific theories has been well documented (Gardner 1981), and generally pseudoscientific claims are aimed at those who do not have sufficient awareness of science or the critical faculties to recognize that they are being fooled. Those who suffer from physical or mental conditions that do not respond to mainstream medicine are particularly susceptible, and there is no doubt that pseudoscientific ideas have widespread credibility in the general population. Indeed, it is in the medical sciences that the boundaries between science and pseudoscience are most confused, with the development and acceptance of complementary medicine opening the door for the legitimization of techniques, such as homeopathy and chiropractic, which are arguably pseudoscientific yet may be prescribed by conventional medical practitioners.

By and large, complementary medicines, remedies and practices do not meet the criteria for evidence-based medicine. This is principally because the claims that are made for their efficacy are not based upon scientifically acceptable rationales (they do not fit into mainstream understandings of human physiology and biology) and because they rely upon anecdotal evidence and personal testimonials rather than randomized clinical trials. Homeopathy, developed in 1796 by the German physician Samuel Hahnemann, and endorsed by the British Royal Family, provides a good case in point. Homeopathy is based upon the so-called "Law of Similars", the principle that ailments will be cured by serially diluted formulations of substances that, in their undiluted form, would cause symptoms similar to those of the ailment for which a cure is being sought. The homeopath assesses the totality of the patient's symptoms and prescribes formulations based on this "holistic" approach to diagnosis and treatment. Homeopathic remedies are prepared by a process called "potentization" in which, at each stage of serial dilution, the formulation is subject to "succussion" by striking on an elastic body. There is no scientific rationale for this process. Furthermore, the serial dilutions of homeopathy (that are claimed to increase in potency as the dilution increases) have been shown to have vanishingly small quantities of the original substance being diluted. The "most potent" homeopathic medicines contain nothing but milk sugar—and the small quantities of contaminants found in all chemical preparations!

Ultimately, the distinction between medical science and pseudoscience centers on the nature of the evidence presented in their support. The gold standard for the validation of a medical treatment, a drug or a medical device, is generally assumed to be a randomized clinical trial. Subjects will be selected using techniques to avoid biases due to medical, ethnic or socio-economic history, and then randomly assigned to treatment groups (for example the drug/test group and a control group). In a double blind trial neither the subjects of the trial nor the researchers themselves will know who is in each group. Well-designed clinical trials will minimize the likelihood that errors will be made due to chance, to placebo effects, or a desire for a positive outcome on behalf of the researchers or their subjects. Such trials will be required for FDA approval of new treatments, and form the basis of evidence-based medicine. However, it should be noted that the status of randomized clinical trials as the "gold standard" is not uncontroversial. It has been claimed that there has sometimes been an uncritical overemphasis on evidence from clinical trials without an appreciation of their limitations, from both statistical and ethical perspectives (Williams 2010).

The defenders of alternative therapies claim that they are not amenable to clinical trials because the approaches that they take are holistic—the patient is treated as a whole, unique individual. It is argued that it would not then be appropriate or feasible to evaluate them in clinical trials because patients in each treatment group in a trial are treated identically. By its very nature, the clinical trial does not allow the alternative therapy to be properly implemented and tested. For this reason, evidence in support of alternative treatments cannot take account of placebo effects, practitioner biases or wishful thinking on behalf of the patient. Of course, such studies, based as they are on anecdote, cannot be independently repeated.

We might feel that, as researchers with a mainstream scientific education, we are immune from such claims. Nevertheless there are fields in which the distinction between science and pseudoscience is by no means clear. A good example is that of parapsychology, the study of the abilities of certain individuals to exhibit psychic phenomena such as precognition or telekinesis. From 1911, laboratory studies of extrasensory perception (ESP) were carried out at Stanford University, and in the 1930s standardized laboratory methods for evaluating studies on ESP were established at Duke University. The studies done at Duke were the subject of substantial academic criticism, principally because of problems with replication of the work. However, in the 1960s and 1970s the scientific community was more open to the investigation of psychic phenomena, and in 1969 the Parapsychological Association became affiliated with the American Association for the Advancement of Science. Despite initial enthusiasm for studies of parapsychology at some mainstream universities, results have generally proved to be inconclusive and, since the 1980s, interest from American public institutions has waned. Assessments of the work done on ESP by parapsychologists have generally concluded that the available evidence is of poor quality and inadequately controlled. A common view of the credible proponents of parapsychological research (such as Carl Sagan and Nobel laureate Brian Josephson) seems to be that, although there is no conclusive scientific evidence of the reality of paranormal phenomena, there seems to be a small number of areas that may well be worthy of continued investigation.

But is parapsychology a pseudoscience? Carl Sagan asserted that extraordinary claims demand extraordinary evidence if they are to be taken seriously, and it is clear that scientific studies to establish the very existence of paranormal phenomena are extraordinarily difficult to conduct. Proponents claim that they are the victims of academic prejudice and the unwillingness of the mainstream research community to accept evidence that threatens current paradigms. It has to be acknowledged that the field of parapsychology has been blighted with fraud, poor experimental design and wishful thinking, but there seems no reason to consider that it is not amenable to the processes of scientific research and critical thinking. We just have to be particularly careful about weighing the evidence.

Ultimately the success of a research program will be dependent upon the degree to which it can demonstrate and explain phenomena, and make testable predictions. Consider the history of phrenology. In the early 19th century Franz Joseph Gall attempted to correlate mental faculties with the shape and size of different areas of the skull, and the field of phrenology achieved a degree of credibility until the end of the century. Phrenology was, from its outset, subject to broad academic skepticism and ultimately sank under the weight of empirical refutation. It may be argued that phrenology was a legitimate research program within the context of the science of the time. It was put forward as a serious scientific theory, and there is no suggestion that Gall and his followers were charlatans. However, we now know that Gall's methods were faulty as judged by modern standards, and that the theory of phrenology did not permit the making of testable, predictions. For these reasons we now consider it to have been a pseudoscience and contemporary attempts to promulgate phrenology are now considered to be pseudoscientific. Many of the features that we see in the history of phrenology have their parallels in the area of parapsychology, except that in the case of the latter the jury is still out.

Finally let us briefly turn to the issue of Intelligent Design (ID), the most recent incarnation of Creationism, which stands in opposition to mainstream theories of evolution such as Natural Selection. Although the debate continues, it is clear that the vast majority of professional scientists and philosophers of biology will argue against the scientific legitimacy of ID despite the attempts of its proponents to argue that it is a valid alternative to Natural Selection and that it should be taught in schools. A powerful assessment of the scientific status of ID was provided on 20 December 2005 by Judge John E. Jones III in the case of Kitzmiller et al. v Dover Area School District et al. in Pennsylvania (Kitzmiller 2005). He declared in his judgment that "ID violates the centuries-old ground rules of science by invoking and permitting supernatural causation....." and that "ID's negative attacks on evolution have been refuted by the scientific community." Furthermore he declared that "it is additionally important to note that ID has failed to gain acceptance in the scientific community, it has not generated peer-reviewed publications, nor has it been the subject of testing and research."

The lack of peer-reviewed publications in the field of ID is a feature of much of pseudoscience, where quotations and references are either misappropriated from mainstream scientific literature, or from studies and anecdotes that have not been though any peer review process. Advocates of pseudoscientific ideas will often appeal to a conspiracy within the academic community. For this reason, test cases conducted in open court, as in the case of ID, set the standard for objective and well-argued review, and provide a reference point and a counter to any arguments against academic conspiracies.

You may ask why the genuine researcher should be interested in the issues of non-science and pseudoscience. It is worth considering that the well being of scientific research and education depends substantially upon funding by government agencies, the good will of elected legislators and upon the confidence of the public. Resources are in short supply and the selection of appropriate areas for funding is vital. It is therefore crucial that those within the scientific community are able to coherently put forward the case for science rather than pseudoscience. It is also important that they are able to argue cogently with those within the scientific community who wish to admit pseudoscientific and wishful thinking into the scientific milieu. This will not be easy. Those who are trying to establish and maintain the standards of science education and research may be condemned as doctrinaire or rigid. They may be arguing against ideas that have a widespread popular appeal and that may promise solutions to some of mankind's most enduring problems. As has been seen in the debate over Intelligent Design, an understanding of the philosophy of science can be vital in winning the argument.

Practical Points

- Familiarize yourself with pseudoscientific literature to sharpen your critical faculties.
- Rigorously apply your critical faculties to your own thinking so that you avoid slipping over into pseudoscientific thinking.
- Beware of confirmatory bias, searching for evidence that confirms our hypotheses and ignoring that which conflicts with them. This is the aspect pseudoscientific thinking that most clearly distinguishes it from a genuine scientific attitude that endeavors to subject hypotheses to rigorous criticism and testing.
- Be especially careful of applying your expertise in one field to another in which you are not an expert. An unfortunate amount of pseudoscience is generated by scientists who are well trained in one area and then apply their knowledge to another.
- Science is under attack from those who believe that truth is relative (anything goes) and from those who reject the critical standards of science in favor of comforting pseudoscientific beliefs. Be prepared to defend science by understanding the philosophy of science so that you can marshal your arguments against pseudoscience.

- Make yourself aware of current controversies on the interfaces between science and metaphysics, and between science and pseudoscience. Contribute to the debate to maintain the acceptance of science as the most effective program for understanding the nature of the material world and for resolving the problems that face us all.

References

Beyerstein BL (2001) Alternative medicine and common errors of reasoning. Acad Med 76:230–237

Feynman R (1996) Six easy pieces. Helix Books. Addison-Wesley Publishing Company. p 4

Gardner M (1981) Science: good, bad, and bogus. Prometheus books. See also http://www.cs icop.org. (The Committee for Skeptical Inquiry publishes the on-line journal The Skeptical Inquirer that uses critical investigation, and the use of reason, to examine controversial and extraordinary claims. Some of the founding members of CSI include scientists, academics, and science writers such as Carl Sagan, Isaac Asimov, Philip Klass, Paul Kurtz, Ray Hyman, James Randi, Martin Gardner, Sidney Hook, and others.)

Kitzmiller v (2005) Dover Area School District. United States District Court for the Middle District of Pennsylvania, 20 Dec 2005

Williams BA (2010) Perils of evidence-based medicine. Perspect Biol Med 53:106–120

Chapter 14
Scientific Thinking: Being Rational?

Abstract The reasoning processes employed in the resolution of research problems, in the analysis and evaluation of scientific data, and in the assessment of experimental results, may be complex. Critical thinking is not something that can be applied mechanistically or simplistically, but is a skill that needs development and practice.

> Let us endeavor then to think well; therein lies the principle of morality.
>
> Blaise Pascal 1623–1662.

We like to think of scientific research as a process that is underpinned by methods of critical thinking that are entirely rational. Karl Popper (1963) proposed that "…there is no more rational procedure than the method of trial and error—of conjecture and refutation: of boldly proposing theories; of trying our best to show that these are erroneous; and of accepting them tentatively if our critical efforts are unsuccessful." The whole panoply of scientific method—generation and testing of hypotheses, experimental design, statistical analyses, the interactive community of researchers, peer review, conferences, and discussion provides the bedrock for the processes of critical reasoning. The community provides a forum for the open exchange and development of ideas while at the same time ensuring, through a process of checks and balances, that the processes of doing research are not subverted. However, when we look at the actual practice of research and the history of science, we see that the processes of critical thinking are not as straightforward as we might at first think, neither in terms of what is seen as constituting a reasoned and rational approach, nor in the ways in which the processes are followed.

The processes of research make use of well-defined logical procedures for deriving conclusions from supporting evidence. In principle we can make deductions from our theories and hypotheses that can be tested experimentally. We can make rational assessments of the level of confidence that we have in competing hypotheses, and we can make reasoned judgments about which of them should

P. Truran, *Practical Applications of the Philosophy of Science*,
SpringerBriefs in Philosophy, DOI: 10.1007/978-3-319-00452-5_14,
© The Author(s) 2013

stand and which should be rejected. However, the reasoning processes employed in the resolution of research problems, in the analysis and evaluation of scientific data, and in the assessment of experimental results, may be complex. Critical thinking is not something that can be applied mechanistically or simplistically, but is a skill that needs development and practice. It is continually in danger of being subverted by mistakes in our reasoning, by our prejudices and biases, by our need to be right, and by our self-interest.

The processes of critically evaluating our theories, and the data that support them, demand an intellectual rigor that can be daunting. One of the first tasks in a research project is to undertake a literature review. This can take the form of a passive gathering of relevant papers and drawing together an overview of work done in the areas of interest. However, it should be an opportunity to make a fair-minded and unbiased assessment of the state of knowledge in the field—a demanding task both in terms of both technical and mental effort. Such a review will allow the researcher to fearlessly examine his views and refine the questions and issues to be addressed. It will enable him to understand how he can make a significant contribution to the existing literature. It will demand a close reading of the literature, both original papers and reviews. It will provide a real understanding of what was done, the reasoning involved and the experimental and statistical procedures used.

There are many potential pitfalls. The reputations of authors, and the journals in which they are published, the rigor of the peer review processes and our respect for the printed word, may tempt us into uncritically accepting what we read. We may place more emphasis on those papers and reviews that endorse our views, rather than those that provide a critique of our position. It takes practice to rapidly identify and assess the validity of the assumptions and premises on which a logical argument is based. Experience will enable the researcher to more effectively review the methods used to generate the data, and to be satisfied that the conclusions presented have been legitimately derived from the data. A thoroughly executed literature review will provide an opportunity for honing the reasoning skills which need to be applied to the researcher's own work.

> ... like any skill, becoming very good at scientific reasoning requires both practice and talent. But becoming tolerably good requires mainly practice and only a little talent. And, for most people tolerably good is good enough. So work at developing your skills little by little.
>
> Ronald N. Giere (1979)

We might think that, although individual researchers may make errors of critical reasoning, the peer review processes, and the systems for review and funding of research proposals provide assurance that good standards of critical thinking will prevail. However, despite their common aim of objectivity, the community of scientific researchers is subject to the same weaknesses and the same dangers of irrationality as all human communities. Researchers work within a paradigm, a "shared set of methods, standards and basic assumptions within a group of research workers in a particular field" and these will determine "...what research is supported, what problems matter, what solutions are admissible, who is promoted, who referees papers, who publishes, who perishes. (Hacking 1983)" The paradigm provides an environment in which education and research can be done without having to continually challenge the fundamentals. It provides a stabilizing effect, and maintains a coherence that allows effective communication within the field. The forces of the paradigm will have a powerful effect on the direction that research takes within a particular field. However, these same forces may sometimes run counter to the canons of scientific rationality, especially in those fields where it is difficult to carry out definitive experiments for either practical or ethical reasons.

The field of obesity, diabetes and nutrition provides an excellent example of this phenomenon. Gary Taubes (2007) has eloquently documented the ways in which 'irrational' forces have shaped the course of research in this field, and the nature of the debate on the best means to address the problem. In the first half of the twentieth century German and Austrian researchers made considerable strides towards the elucidation of the links between insulin, obesity, and diabetes. Studies on genetically obese mice showed that they gained weight even when they were prevented from overeating. They proposed that obesity should not be explained simply as a phenomenon due to caloric imbalance (greater input than expenditure). Rather, they maintained that fatty tissues store lipids regardless of energy balance. Obesity was seen to be a disorder of metabolism and hormonal balance. However, after the Second World War, antipathy towards Germany, and the rise of English as the primary language of science, meant that much of this work was ignored. The German literature on obesity rapidly disappeared from the field. At the end of the war, those German scientists in the field of obesity and nutritional research who did make it to the United States ended up in minor universities where they had little impact. As a result, the alternative hypothesis that obesity was due to overeating (excess caloric intake) began to dominate the field, buoyed up by developments in psychologically-determined behavioral approaches. The overlooking of German research, and its effect on post-war research on obesity, was entirely understandable in human terms. It was irrational in scientific terms.

Although research on insulin and fat metabolism made great strides in the postwar years, it had little impact on mainstream thinking about obesity. The hypothesis that dietary fat was a major component in the elevation of blood cholesterol level and the development of heart disease became widely accepted. This drove established opinion away from the consideration of high carbohydrate consumption, with its impact on the insulin-modulated balance between glucose consumption and fat deposition, as the principal cause of obesity. Instead, the reduction in fat consumption and a reduction of caloric intake, were perceived as the recommended route to a healthy diet. Taubes has shown how established wisdom in the study of nutrition, chronic disease, and obesity was determined by a small number of influential experts who had a disproportionate effect on national nutritional policy. In 1977 the US Government published "Dietary Goals for the United States", which recommended caloric proportions of 55–60% carb, 30% fat (no more than a third of that saturated). The establishment of a government position on dietary fat turned a scientific position into a political issue, and the recommendations were enshrined in federal policy. The politics were further complicated by the funding of a number of dietary research institutions by the food industry, which had commercial interests in the shift towards an increased consumption of carbohydrates rather fats. The policy was maintained despite the lack of any definitive evidence from well-defined experimental trials. Confirmation bias was common. Consideration was given only to supporting evidence, and data from trials that cast doubts on the dietary fat hypotheses were dismissed. Once the perceived benefits of a low fat diet had been established in the domain of public opinion, it became almost impossible for the problem to be studied in a rational and scientific manner.

> ... it is the peculiar and perpetual error of the human intellect to be more moved and excited by affirmatives than by negatives; whereas it ought properly to hold itself indifferently disposed toward both alike.
>
> Francis Bacon, Novum Organum (1620)

It is evident that the problem of establishing the relationship between obesity, diabetes and diet is a complex one. However, the public and political pressures to come up with answers and policies, as well as the temptations of personal or corporate advantage, militated against doing good science. The general public often finds it hard to accept that the results of scientific research are often provisional, wrapped around with estimates of probability and with significant caveats. There was a demand for clear directives in the face of an explosion of obesity and diabetes. Most importantly, these pressures made it difficult for leaders in the field to admit to the uncertainties about the solutions to the problem of obesity. The power of the paradigm ensured for decades that dissenting voices did not significantly influence the direction of research within the field and the recommendations that were made.

The rational attitude demands humility and courage in the face of the real world. Sometimes an idea, such as the reduction of dietary fat to reduce of heart disease and obesity, may be so obvious or make so much sense that we feel that it just has to be true. As evidence mounts that a dearly held hypothesis may not be correct, it may be difficult to resist the temptations of confirmation bias or even to fudge, select or, sometimes, fabricate data. At this point it is vital to remember that all successful researchers have, at some point, gone down the wrong track. Any element of our knowledge can turn out to be mistaken, and it is by recognizing our mistakes that we can change direction and make progress. Even so, this can be difficult and dispiriting. It is heartening, therefore, to read accounts like that of the study carried out at the University of Cincinnati, College of Medicine to assess the effectiveness of bypass operations to improve blood flow to the brain. It seemed intuitive to these researchers that increasing blood flow by bypassing blocked carotid arteries would reduce the incidence of stroke in patients. However, the results of the study were hugely disappointing. Patients who received costly bypass operations had no significant reduction in the incidence of stroke in the 2 years following surgery when compared with patients who received only medication. In 2011 it was reported (Powers et al. 2011; Grady 2011) that the entire multi-million dollar program had been abandoned. Dr. William J. Powers, the lead author of the study said "I've probably put 30 years of my life into this question," and he added that stopping the study was the right decision "if there is literally no chance we're going show the surgery works." It takes courage and integrity to face up to disappointments like this. It is a harsh reality that scientific kudos is associated with the successful solution of problems, rather than from the identification of areas of research that are no longer worth pursuing. The acceptance of uncomfortable and inconvenient truths that we may discover about the real world truly reflects a rational scientific attitude.

The exercise of reasoning and critical thinking requires constant vigilance and practice. It will be rooted in an intellectual humility that recognizes the limits of one's personal knowledge, and the uncertainties that pertain in our field of research. It understands that our knowledge and our thinking are prone to error, that we make mistakes, and that our work may be distorted by our prejudices, bias and our self-interest. In this respect the exercise of reason is rooted in a moral integrity that places the search for the truth above personal ambition. This moral integrity is distinguished by a respect for intellectual rigor, commitment and fairness. The fledgling researcher may not consider that these larger issues of the pressures of the paradigm or institutional of inertia on the direction of a research program are of immediate importance. However, the forces of irrationality may be ever present, in the library, in the laboratory and in the conference room. It is not easy to be rational, but the willingness and humility to acknowledge the deficiencies in our knowledge provides the bedrock for scientific rationality and critical thinking.

Practical Points

- Develop an understanding of your capabilities for critical thinking, and become aware of your weaknesses. Learn to be an effective critic of your own reasoning so that you recognize when it is muddled, confused, inaccurate and illogical. Be willing to rigorously critique your ideas so that you can refine them to the point where they are worth testing and validating.
- Develop an intellectual humility—an understanding and acceptance of the limits of your knowledge, and the real uncertainties that attend it. Intellectual humility is the opposite of intellectual arrogance. Be aware of your prejudices and biases, and do not claim more than you actually know.
- Ensure that the assumptions that you make in support of an argument are supported by evidence. In research it is easy to make unwarranted assumptions about the appropriateness, performance and validation of the instruments and methodologies that we apply in our work.

- Embrace uncertainty. At a fundamental level we can consider all knowledge to be, in principle, provisional or uncertain. In practice, we do not challenge the bedrock of our understanding of the world, but if we embrace uncertainty we open ourselves up to the possibilities that we may be mistaken and that the world may be seen in a new way.
- Be open to new ideas, and be prepared to reject those that do not stand up to critical scrutiny.
- Recognize that knowledge is subject to ambiguity, and that we need to make our language as clear as possible. This does not mean that we should pay inordinate attention to definitions, but we should endeavor to present our arguments in terms that are understood by those to whom they are directed. Be prepared to clarify, define and get agreement on any terms whose meaning may not be self-evident. Use examples or illustrations to make your position clear.
- Cultivate intellectual independence. Be prepared to challenge the status quo, but take account of when and how you do this. Be aware of the dangers.
- Be alert to the assessment of factual claims, particularly with respect to their accuracy and precision. Remember that all facts have theoretical or hypothetical components that may need to be made explicit or clarified.
- Scientific reasoning is a skill that needs practice. Educate yourself about the rules and applications of logic and scientific reasoning. There are plenty of excellent guides and a number of them are listed below (Giere 1979; Moore and Parker 2009; Paul and Elder 2002).

References

Giere RN (1979) Understanding scientific reasoning. Holt, Rinehart and Winston, New York

Grady D (2011) Study debunks operation to prevent strokes. New York Times, New York

Hacking I (1983) Representing and intervening. Cambridge University Press, Cambridge

Moore BN, Parker R (2009) Critical thinking, 9th edn. McGraw Hill, New York

Paul R, Elder L (2002) Critical thinking: taking charge of your professional and personal life. Financial Times Prentice Hall, Harlow

Popper KR (1963) Conjectures and refutations. Routledge and Kegan Paul, London

Powers WJ, Clarke WR, Grubb RL, Videen TO, Adams HP, Derdeyn CP (2011) Extracranial-intracranial bypass surgery for stroke prevention in hemodynamic cerebral ischemia: the carotid occlusion surgery study randomized trial. JAMA: J Am Med Assoc 306

Taubes G (2007) Good calories, bad calories. Anchor Books, New York

Chapter 15
Some Concluding Thoughts

Abstract All our knowledge is hypothetical and provisional. However, we need not be pessimistic about the status of our knowledge. Although our explanations require the agreement of a complex network of theories and observations, we can still confidently trace the patterns of evidence that provide the justifications for our beliefs. Such beliefs provide a powerful basis for understanding and manipulating the world.

The aim of this book was to demonstrate that a consideration of the key principles of the philosophy of science can be of practical help in doing research. While it is evident that no single view can adequately capture the nature of science as a distinctive activity, successful research seems to depend as much upon the adoption of appropriate *attitudes* towards knowledge as upon the application of logically rigorous methodologies. Furthermore, these attitudes and methodologies may be appropriate within fields of knowledge that are not traditionally seen as scientific.

An understanding of these principles can be helpful in any situation where it is necessary to systematically resolve problems. They provide guidance for research in the physical and biological sciences, but are also invaluable in the applied sciences and technology, as well as in engineering and design. Indeed, we can use them in any area where we can apply logic, reason and structured investigation. The attitudes outlined here will sharpen our ability to assess the value of new ideas. Rigorous appraisal and critique will alert us to internal weaknesses or potential alternatives. It will help us to recognize the importance of careful, robust experimental designs that take into account biases that might erroneously support a particular outcome. The conscientious researcher will therefore become familiar with the developments that have taken place in the design of experiments and in the use of statistics for the interpretation of data. Recognition of these principles will alert him to the potential for error and, in particular, for assumptions that may confound the validity of experiments, observations and conclusions.

P. Truran, *Practical Applications of the Philosophy of Science*,
SpringerBriefs in Philosophy, DOI: 10.1007/978-3-319-00452-5_15,
© The Author(s) 2013

Perhaps the most important element of the scientific attitude is the acceptance that all our knowledge is hypothetical and provisional. In principle, nothing is certain. However, the notion that there is no ultimate justification for our beliefs about the world does not mean that such beliefs are arbitrary. We need not be pessimistic about the status of our knowledge. Although the general acceptance of our explanations of phenomena requires the agreement of a complex network of theories and observations, we can still confidently trace the patterns of evidence that provide workable and acceptable justifications for our beliefs. Such beliefs provide a powerful basis for understanding and manipulating the world.

The confidence we have in the results of scientific research will depend upon the degree of integrity we perceive in the research community. We cannot repeat every finding reported in the research literature. Much has to be taken on trust, just as we read a leading newspaper with a confidence generated from years of reading it. We recognize the newspaper's acknowledgement and correction of errors of fact, its publication of letters expressing contrary views and the columns written by journalists with differing political or economic viewpoints. At root, we know that if we really did want to verify a report we probably could do so. We could visit the site of the report and speak to witnesses ourselves; or we could check the credentials and reputation of the journalist who filed the report. Similarly, our trust in the findings of scientific research depends upon the confidence we have in the great scientific institutions, research establishments and journals. It will be strengthened by the application of reason and logic to our ideas, and their open review and critique by our scientific peers. This confidence begins to weaken when the systems of checks and balances provided by peer review, by financial independence and by freedom from political, religious or ideological pressures are compromised.

Our understanding of the philosophy of science helps us to develop the attitudes and rigor that are required for the expansion of scientific knowledge by effective research. It also helps us to appreciate the role we have in understanding the world, and the means we have for changing and bettering it.

Further Reading

Hopefully you will have been persuaded to read more on the philosophy of science, and on critical thinking. There is a wide literature that discusses the philosophy of science from a layman's perspective and also from a standpoint that assumes a professional knowledge of the field. Within the physical and biological sciences, a number of researchers have considered their fields from a philosophical perspective and provide clear, practical advice on how to do research. A few of the most useful are listed below.

Committee on the Conduct of Science, National Academy of Sciences (1989) On being a scientist. Proc Natl Acad Sci USA 86:9053

Feibelman PJ (1993) A Ph.D. is not enough: a guide to survival in science. Addison-Wesley, Reading, MA

Giere RN (1979) Understanding scientific reasoning. Holt, Rinehart and Winston, New York

Grinnell F (1992) The scientific attitude, 2nd edn. Guilford Press, New York

Grinnell F (2009) Everyday practice of science: where intuition and passion meet objectivity and logic. Oxford University Press, Oxford

Medawar P (1979) Advice to a young scientist. Alfred P. Sloan Foundation Series

Medawar P (1969) Induction and intuition in scientific thought. American Philosophical Society, Philadelphia

Moore BN, Parker R (2009) Critical thinking, 9th edn. McGraw Hill, New York

Index

P. Truran, *Practical Applications of the Philosophy of Science*,
SpringerBriefs in Philosophy, DOI: 10.1007/978-3-319-00452-5,
© The Author(s) 2013